AI-Driven Security for Next-Generation IoT Systems

Mourade Azrour · Abdulatif Alabdulatif
Editors

AI-Driven Security for Next-Generation IoT Systems

Springer

Editors
Mourade Azrour
Department of Computer
Faculty of Sciences and Technologies
Moulay Ismail University of Meknès
Errachidia, Morocco

Abdulatif Alabdulatif
Department of Computer Science
College of Computer
Qassim University
Buraydah, Saudi Arabia

ISBN 978-3-032-08783-6 ISBN 978-3-032-08784-3 (eBook)
https://doi.org/10.1007/978-3-032-08784-3

This Springer imprint is published by the registered company Springer Nature Switzerland AG
The registered company address is: Gewerbestrasse 11, 6330 Cham, Switzerland

Preface

The rapid expansion of Internet of Things (IoT) devices and systems has transformed our quotidian activities in various areas, from agriculture, healthcare and transportation to finance and smart infrastructures. For instance, these interconnected devices become more central to everyday existence, things that implicate the importance of ensuring their security strengthens. The present book, titled AI-Driven Security for Next-Generation IoT Systems, explores a comprehensive collection of cutting-edge techniques and strategies aimed at protecting these complex networks. Furthermore, it emphasizes the shift from traditional security equipment to sophisticated, vulnerability-managed frameworks that can adapt flexibly to emerging threats. The chapters collectively highlight the requirement of harnessing advanced artificial intelligence (AI) and machine learning (ML) approaches to address the evolving landscape of IoT threats, underscoring the intersection of security, scalability, and automation in next-generation IoT environments.

Subsequently, the chapters investigate the fundamental methodologies and innovations transforming IoT security. Hence, topics treated range from the assessment of deep learning approaches for intrusion detection to the development of multi-factor authentication schemes based on elliptic curve cryptography. These sections investigate how AI-driven algorithms—such as deep Q-networks and ensemble learning—are being utilized for real-time threat detection, anomaly recognition, and automated response mechanisms. Moreover, a particular focus is given to the integration of dynamic computational intelligence algorithms and hybrid deep learning models, which enhance the resilience and precision of intrusion detection systems across various IoT and edge computing frameworks. Additionally, the book explores the application of blockchain technology, graph neural networks, and transfer learning to safeguard sensitive data, verify transactions, and improve biometric authentication within complex, distributed networks.

Finally, the book underlines the importance of comprehensive security frameworks that combine innovative cryptographic protocols, intelligent analytics, and secure communication mechanisms. These include PUF-based authentication for drone communication, RFID mutual authentication schemes, and AI-supported digital forensics tailored to the Internet of Medical Things (IoMT). By systematically analyzing these advanced mechanisms, the chapters aim to equip researchers, practitioners, and policymakers with practical insights into designing resilient, scalable, and intelligent security architectures. The collective knowledge presented here paves the way for future innovations in AI-driven cybersecurity, ensuring that next-generation IoT systems can function securely, efficiently, and confidently within an increasingly connected world.

Errachidia, Morocco Mourade Azrour
Buraydah, Saudi Arabia Abdulatif Alabdulatif

Contents

From Security Equipment to Managing Vulnerabilities

Yasmine EL Barhami and Hatim Kharraz Aroussi

Abstract As cyber threats grow more advanced, relying solely on security equipment is no longer enough to protect organizations effectively. In this paper, we will discuss the limitations of traditional cybersecurity approaches and emphasize the importance of incorporating continuous monitoring and risk assessment into broader security strategies. Drawing from real-world case studies, we will demonstrate how proactive steps such as regular audits and employee training, can significantly reduce the risk of cyberattacks. While tools like firewalls, intrusion detection systems (IDS), and endpoint detection and response (EDR) remain critical, many organizations still experience serious breaches. This leads to an essential question: **Can security tools alone keep up with today's evolving threats?** Often, the root cause of breaches lies in vulnerabilities, whether due to outdated software, hardware flaws, or misconfigurations. Security tools can't fully address these issues without a structured vulnerability management process in place. Our analysis underscores the need to move beyond a tool-based mindset and adopt a comprehensive vulnerability management approach. This shift enables organizations to detect, prioritize, and remediate risks more effectively, resulting in a stronger and more adaptive cybersecurity posture. When done well, vulnerability management closes security gaps before they're exploited. By encouraging a culture of proactive defense, businesses not only boost their cyber resilience but also set themselves up for long-term success in an increasingly digital landscape.

Keywords Cybersecurity · Vulnerability management · Security equipment · Risk assessment · Continuous monitoring · Employee training

Y. EL Barhami (✉) · H. K. Aroussi
Ibnu Tofail University, Kenitra, Morocco
e-mail: Yasmine.elb@gmail.com

H. K. Aroussi
e-mail: hatim.kharrazaroussi@uit.ac.ma

M. Azrour and A. Alabdulatif (eds.), *AI-Driven Security for Next-Generation IoT Systems*, https://doi.org/10.1007/978-3-032-08784-3_1

1 Introduction

In today's digital landscape, organizations face an increasingly complex array of cybersecurity challenges. The evolving threat put significant risks, as cybercriminals continuously develop sophisticated tactics such as ransomware, phishing, and advanced persistent threats. These attacks are becoming more targeted and damaging, forcing organizations to remain vigilant and adaptive in their defenses.

The proliferation of remote work, Internet of Things (IoT) devices, and cloud computing has further expanded the attack surface [1–4]. Each new device or platform introduces potential vulnerabilities that attackers can exploit, making it imperative for organizations to monitor and secure all endpoints effectively. This complexity is compounded by the diverse array of security tools and technologies that organizations must manage, often leading to gaps in security posture if not properly integrated.

Human error remains a significant factor in cybersecurity breaches. Poor password practices, inadequate training, and susceptibility to social engineering attacks can undermine even the most robust security measures. It is crucial for organizations to invest in employee training and awareness programs to mitigate these risks.

Additionally, compliance with various regulations presents its own set of challenges. Organizations must implement robust policies and practices to meet these requirements, as non-compliance can result in severe penalties and reputational damage.

Finally, resource constraints often hinder effective cybersecurity strategies. Many organizations struggle with limited budgets, personnel, and expertise, making it difficult to implement comprehensive security measures. Given these challenges, effective vulnerability management emerges as a critical component of a proactive cybersecurity strategy, enabling organizations to enhance their resilience against emerging threats [5, 6]

2 Research Methodology and Selection Criteria

Our study adopts a qualitative and analytical approach, examining industry's best practices, real-world case studies, and existing research on cybersecurity strategies to analyze the transition from security equipment to effective vulnerability management. The methodology consists of:

Literature Review: We reviewed academic papers, industry reports, and cybersecurity frameworks (such as NIST, ISO 27001, and CIS) to understand the evolution of security approaches from hardware-based defenses to proactive vulnerability management.

Case Studies and Incident Analysis: High-profile cyber incidents were examined to highlight how security breaches occurred despite organizations investing in traditional security equipment. These cases are explained in Sects. 3.1.1, 3.1.2, 3.1.3.

Selection Criteria: Our study focused on organizations from diverse industries, including finance, healthcare, and technology, that have either suffered security breaches due to poor vulnerability management or successfully adopted a proactive approach. The key selection criteria included:

- Organizations with publicly documented security incidents related to unpatched vulnerabilities.
- Studies and reports published in reputable cybersecurity journals and whitepapers such as Google Scholar, Scopus..., etc.

By applying this methodology, our article aims to provide a perspective on why organizations must go beyond traditional security equipment and adopt a continuous vulnerability management process.

3 The Role of Security Equipments

3.1 Types of Security Tools

Security tools are essential components of an organization's cybersecurity strategy, each serving for a specific purpose in protecting against threats.

The famous Firewall act as a barrier between a trusted internal network and untrusted external networks. It filters incoming and outgoing traffic based on predetermined security rules, blocking unauthorized access while allowing legitimate communications. The Intrusion Detection Systems (IDS) monitor network traffic for suspicious activities and potential threats [7–11]. They analyze patterns and can alert administrators to potential breaches, enabling a timely response. IDS can be network-based or host-based, depending on their deployment. Similar to IDS, the Intrusion Prevention Systems (IPS) not only detect threats but also take action to prevent them. They can automatically block malicious traffic, making them a more proactive defense mechanism.

Also, there are Antivirus and Anti-malware Software that can scan and remove malicious softwares from endpoints, protecting against viruses, spyware, and other forms of malwares. Regular updates are crucial to keep these tools effective against new threats. For sensitive data protection we can talk about encryption tools by converting it into a format that is unreadable without a decryption key. This ensures that even if data is intercepted, it remains secure.

Then, come the Security Information and Event Management (SIEM): SIEM systems aggregate and analyze security data from various sources, providing real-time insights and alerts [11–14].

Each of these techniques and tools plays a vital role in an organization's cybersecurity framework, but they focus only on detecting, blocking threats and controlling access but they cannot address underlying vulnerabilities, such as outdated software or misconfigurations, that attackers can exploit. Without a structured approach to identify, assess, and remediate these weaknesses, organizations risk being overwhelmed by alerts and unable to respond effectively. Besides, integrating vulnerability management with security tools is crucial for creating a resilient defense, significantly reducing risk exposure and enhancing overall cybersecurity effectiveness [15–17]

Let's look at a comparative analysis of traditional security equipment approaches with vulnerability management frameworks in Table 1.

Table 1 Comparative analysis of traditional security equipment approaches with vulnerability management frameworks

Aspect	Traditional security equipment approaches	Vulnerability management frameworks
Purpose	Protect systems using hardware and software tools like firewalls, IDS/IPS	Identifies, assesses, and remediates vulnerabilities in systems and software
Approach	Preventive and reactive security using perimeter defenses and endpoint controls	Proactive and continuous assessment of security weaknesses across the environment
Focus	Securing network boundaries and endpoints (e.g., firewalls, anti-virus)	Identifying weaknesses in software, configurations, and infrastructure to minimize risk
Scope	Primarily focuses on external threats and blocking access	Comprehensive, focusing on internal and external vulnerabilities across all assets
Tools	Firewalls, IDS/IPS, Anti-virus, VPNs, Encryption, Access Controls	Vulnerability scanners, patch management, risk assessment tools, continuous monitoring
Response Time	Typically reactive: responds after an event or detected attack	More proactive: continuous monitoring and patches vulnerabilities before exploitation
Visibility	Provides visibility over network traffic, users, and external threats	Provides deep insight into system vulnerabilities, configurations, and patch statuses
Complexity	It can be complex in terms of maintaining hardware, firewalls, and other controls	Requires ongoing vulnerability scans, analysis, and patching with evolving threat intelligence
Cost	High upfront cost for equipment and maintenance	Can be cost-effective over time, relying on software tools and ongoing vulnerability management
Flexibility	Generally fixed equipment setups, with limited scalability	Scalable and adaptable to dynamic environments and continuously evolving threats
Effectiveness	Effective against known attacks but limited against zero-day or emerging threats	Highly effective in identifying and remediating vulnerabilities before they are exploited

3.2 *The Story of Vulnerabilities*

The concept of vulnerabilities in cybersecurity is not new; it has evolved alongside the development of technology. From the early days of computing in the 1960s, when systems were isolated and primarily academic, the notion of security focused on protecting physical hardware and maintaining access control. However, as networks expanded in the 1980s and the internet began to emerge, the complexity of systems increased, exposing new weaknesses or flaws in a system, application, or network that can be exploited by attackers to gain unauthorized access or cause harm. They can arise from various sources, including software bugs, misconfigurations, or human error. In the 1990s, high-profile incidents, such as the Morris Worm, highlighted the consequences of software vulnerabilities, leading to greater awareness and the need for structured approaches to security. This period saw the birth of vulnerability databases and the development of standards aimed at identifying and mitigating weaknesses. As the 2000s progressed, cyber threats grew more sophisticated, with vulnerabilities becoming central to exploit strategies. The establishment of organizations like the Common Vulnerabilities and Exposures (CVE) in 1999 marked a pivotal moment, creating a standardized way to catalog and discuss vulnerabilities across the industry [18]

Understanding their historical evolution provides valuable insights into how organizations can better protect themselves in an ever-evolving digital landscape.

After getting this, we need to understand also the types of vulnerabilities that we are talking about in this article, we can start with:

- **Software Vulnerabilities**: Flaws in applications or operating systems that can be exploited, such as buffer overflows or unpatched software.
- **Configuration Vulnerabilities**: Incorrect settings in hardware or software that can expose systems to attacks, such as default passwords or open ports.
- **Network Vulnerabilities**: Weaknesses in network protocols or architecture, such as unsecured Wi-Fi networks or lack of segmentation.
- **Human Vulnerabilities**: Mistakes or poor practices by users, such as weak passwords or falling for phishing attacks.
- **Physical Vulnerabilities**: Risks related to physical access to systems, such as inadequate security measures at data centers or unauthorized personnel access.
- **Supply Chain Vulnerabilities**: Weaknesses introduced through third-party vendors or partners, which can compromise the security of an organization's systems.

Understanding vulnerabilities and their types lays the foundation for effective vulnerability management. Once an organization identifies and categorizes its vulnerabilities, the next critical step is to implement a structured vulnerability management process. This process ensures that vulnerabilities are not only recognized but also systematically addressed to minimize risk.

By following a series of defined steps, organizations can prioritize vulnerabilities based on their potential impact and likelihood of exploitation. This proactive approach enables teams to develop strategies for remediation, ensuring that security measures evolve alongside emerging threats [19, 20].

In the next paragraph, we will focus more on the steps of managing vulnerabilities which is a crucial aspect of maintaining the security and integrity of any system. Managing vulnerabilities involves identifying, assessing, and prioritizing potential weaknesses in a system, and then taking proactive measures to address them.

3.3 *Steps of Vulnerability Management Process*

The vulnerability management lifecycle has five key phases. Organizations looking to implement or improve their vulnerability management program can follow these steps. The following steps as shown in Fig. 1 outline a robust vulnerability management strategy:

- **Identification:** Discover and catalog all assets, including hardware, software, and network components. Use vulnerability scanning tools to identify known vulnerabilities in these assets.
- **Assessment**: Evaluate the identified vulnerabilities to determine their severity and potential impact. This often involves using frameworks like Common Vulnerability Scoring System (CVSS) to prioritize them based on risk.

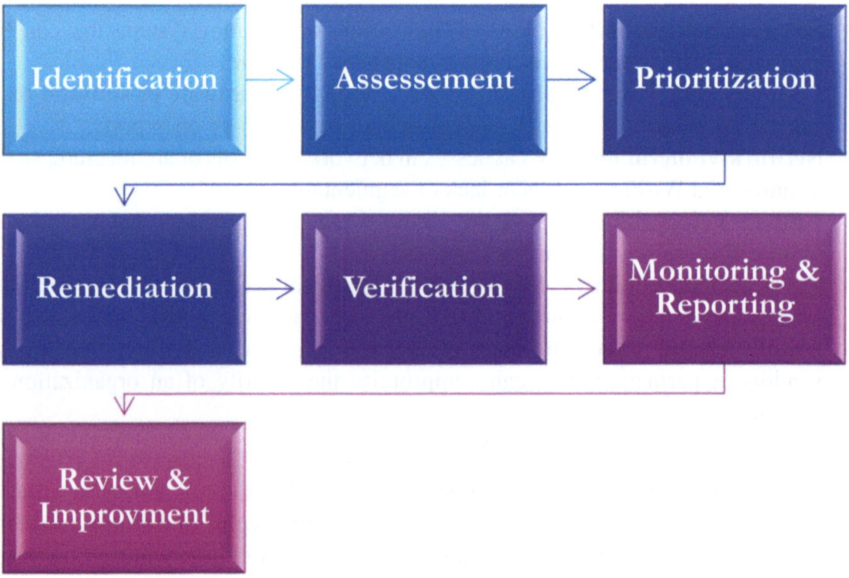

Fig. 1 Vulnerability management process

- **Prioritization**: Rank vulnerabilities according to their risk level, potential impact on business operations, and exploitability. Focus on addressing the most critical vulnerabilities first.
- **Remediation**:
  - Automatic remediation: The use of automated remediation approaches has become more common as they allow for rapid identification and patching of identified threats. The tools and procedures used in automated remediation often follow predefined rules and processes, which allow for consistency and minimal risks due to human error during the remediation process
  - Manual remediation: Develop and implement strategies to mitigate or eliminate vulnerabilities. This may involve applying patches, changing configurations, or implementing additional security controls.
- **Verification**: After remediation, verify that vulnerabilities have been successfully addressed. Conduct follow-up scans and tests to ensure that the applied fixes are effective.
- **Monitoring and Reporting**: Continuously monitor the environment for new vulnerabilities and changes. Regularly report on vulnerability status and management efforts to stakeholders to maintain transparency and accountability.
- **Review and Improvement**: Periodically review the vulnerability management process to identify areas for improvement. Incorporate lessons learned and adapt strategies based on evolving threats and organizational needs.

Having process is not enough to handle all the vulnerabilities, they remain to have some tools that help the organization in the management. The crucial tools are vulnerabilities scanner, Patch management systems, SIEM and pentesting tools. The integration of these tools into the vulnerability management process enhances overall cybersecurity resilience, enabling organizations to proactively defend against potential attacks.

4 Cyberattacks in 2024

4.1 Analysis of Organizations that Failed Due to Poor Vulnerability Management

In 2024, several cybersecurity incidents underline the importance of robust vulnerability management. Organizations that failed to implement proactive security measures, conduct regular vulnerability assessments, or swiftly remediate known weaknesses found themselves at the center of devastating breaches. From ransomware attacks crippling major enterprises to data leaks exposing sensitive customer information, these incidents highlighted the consequences of poor or nonexistent vulnerability management. Let's examine in this article some of the most impactful breaches of the 2024 year.

4.2 UnitedHealth Attack (Ransomware)

In February 2024, UnitedHealth Group's subsidiary, Change Healthcare, experienced a significant ransomware attack attributed to the ALPHV/BlackCat group. They were able to access UnitedHealth's systems through a server that was not protected with Multi-Factor Authentication (MFA).Threat actors were able to access the private health information of UnitedHealth customers. This breach compromised the personal and healthcare data of approximately 190 million individuals, making it one of the largest healthcare data breaches in U.S. history.

UnitedHealth paid a $22 million ransom to regain access to their systems, but it is estimated that the total cost of the breach will be $1.4–1.6 billion.

This attack shows to importance of using cybersecurity tools that are available and also adopting vulnerability management tool that could identify that the authentication in the server did not have the MFA enabled which is vulnerable as high or critical severity.

In that case, the security team would consider this vulnerability to be remediate as soon as possible so avoid such a compromise [21, 22]

4.3 Ticketmaster Breach (Data Breach)

In May 2024, Ticketmaster, a leading ticket sales and distribution company, experienced a significant data breach that compromised the personal information of approximately 560 million customers worldwide.

The breach occurred between April 2 and May 18, 2024, when an unauthorized third-party accessed a cloud database hosted by a third-party data services provider. The compromised data included:

- **Personal Information**: Names, email addresses, phone numbers, mailing addresses, and ticket purchase history.
- **Payment Details**: Partial credit card information, such as the last four digits and expiration dates.
- **Additional Data**: Customer fraud details and, in some cases, event barcodes for over 70,000 events, including high-profile concerts like Taylor Swift's Eras Tour.

The data breach was at the hands of the hacking group ShinyHunters who sell the information's on the Dark Web. The public found out about this breach on May 28, when information about the data for sale was posted on a message board.

Shiny Hunters group accessed Ticket master through their Snowflake account, a cloud data storage and analysis company. It has been suggested that they installed a Trojan horse virus on the computer of an employee of a different company that also used Snowflake and were able to maneuver into Ticketmaster from there.

As organizations become more interconnected, threat actors have more vectors through which to launch their attacks. A supply chain is only as strong as its weakest link, and threat actors are working tirelessly to find those vulnerabilities.

Once again, this attack was made easier by a lack of MFA. Again, if there was a vulnerability management tool such as vulnerabilities scanner or pentesting tool, this vulnerability will be identified during regular scan as critical or high depending on the type of server. The security team would take appropriate actions to remediate it by enabling MFA [23].

4.4 CrowdStrike (IT Outage)

In July 2024, CrowdStrike, a leading cybersecurity firm, experienced a significant incident involving a faulty software update that disrupted millions of Windows computers worldwide. On July 19, 2024, CrowdStrike released an update to its Falcon endpoint detection and response agent. This update contained a defective configuration file that caused approximately 8.5 million Windows devices to crash, leading to widespread system outages across various industries, including airlines, broadcasters, banks, hospitals, and retailers.

The faulty update resulted in critical disruptions:

- **Airlines**: Delta Air Lines reported over 6000 flight cancelations due to the outage, estimating losses of approximately $500 million.
- **Broadcasters**: Sky News and other media outlets experienced temporary service outages.
- **Financial Institutions**: Banks faced operational challenges, affecting customer transactions, and services.
- **Healthcare Services**: Hospitals encountered difficulties in accessing critical systems, impacting patient care.

CrowdStrike acknowledged the issue and launched a patch to rectify the problem. However, affected devices were unable to connect to the internet to download the patch due to the Falcon agent's crash. The recommended solution involved booting devices into safe mode or Windows Recovery Mode to manually delete the problematic configuration file. This process required local administrator access and, in cases of device encryption, the BitLocker recovery key [24–26].

The CrowdStrike IT outage can be explained as a failure in vulnerability management with the lack of proactive risk assessment and controlled update deployment. If there were a vulnerability management tool, after a regular scan it would show that the software concerned is vulnerable and need to be updated.

4.5 Successful Implementations of Vulnerabilities Management

By adopting comprehensive vulnerability management frameworks that prioritize timely identification, assessment, and remediation, these organizations have fortified their defenses against emerging threats. This transition emphasizes the critical importance of learning from past failures to develop effective strategies that ensure resilience and security in today's complex cyber landscape.

Here are some real cases that have successful implementations stories:

4.6 Microsoft's Patch Management Strategy

Microsoft has developed a robust patch management process that is integral to its cyber security strategy. The company regularly releases security updates and patches on "Patch Tuesday," which occurs on the second Tuesday of each month. They utilize automated tools such as:

- Microsoft Defender for Endpoint
- Microsoft Defender Vulnerability Management (MDVM)
- Windows Update Services (WSUS)
- System Center Configuration Manager (SCCM)
- Azure Security Center
- PowerShell DSC (Desired State Configuration)
- Microsoft Sentinel (SIEM)
- Azure DevOps (with Integrated Security Scanning).

This proactive approach has significantly reduced the number of vulnerabilities that could be exploited by attackers. Microsoft's commitment to timely updates has enhanced the security of its software products, contributing to the company's reputation as a leader in cybersecurity.

4.7 Cisco System

Cisco has implemented a multi-layered vulnerability management strategy that includes:

- Automated scanning and patch management: Cisco utilizes internal tools such as: Cisco Security Manager provides that centralized management for firewalls, intrusion prevention systems (IPS), VPNs, and other security devices, it helps automate the configuration, monitoring, and patch management of security devices. Cisco Identity Services Engine (ISE) helps automate network access control,

policy enforcement, and vulnerability management to continuously monitor its infrastructure.
- Cisco SecureX and Talos Intelligence: These platforms provide real-time threat intelligence, helping Cisco prioritize vulnerabilities based on risk level.
- Cisco Bug Bounty Program: Cisco actively engages security researchers through a bug bounty program to discover and fix vulnerabilities before they can be exploited.
- Zero Trust Architecture: By implementing Zero Trust principles, Cisco minimizes exposure to potential threats.
- Rapid Incident Response: Their Security Operations Center (SOC) ensures that vulnerabilities are patched quickly, reducing the risk of exploitation.

4.8 Google—BeyondCorp and Chrome Vulnerability Management

Google's BeyondCorp model represents a paradigm shift in cybersecurity by eliminating traditional VPN dependencies and enforcing a Zero Trust framework, ensuring that access to company resources is continuously verified based on multiple security signals rather than assumed to be safe within a network perimeter. This model enhances security by treating every access request as potentially untrusted, regardless of whether it originates from inside or outside the organization. In addition to this robust approach, Google maintains strict vulnerability disclosure policies, encouraging transparency and rapid mitigation of security flaws. Through its Google Vulnerability Reward Program, the company actively incentivizes security researchers to identify and report vulnerabilities, further strengthening its security ecosystem. Furthermore, Google prioritizes proactive security measures by ensuring that Chrome browsers receive automatic security updates, reducing user exposure to web-based threats such as phishing attacks, malicious extensions, and zero-day exploits. By combining Zero Trust principles with proactive vulnerability management and continuous security enhancements, Google sets a strong example for how organizations can fortify their digital environments against evolving cyber threats.

5 Limited Visibility into Vulnerabilities

One of the significant limitations of vulnerability management is the lack of visibility into vulnerabilities. With the increasing complexity of modern systems and networks, it has become challenging to identify all vulnerabilities. According to a report, the average organization has over 150,000 vulnerabilities in their systems and networks, but only 20% of them are remediated (Source: Veracode). This lack of visibility makes it difficult for organizations to prioritize and remediate vulnerabilities effectively. Another challenge is the emergence of new vulnerabilities every day. As new technologies and software are introduced, new vulnerabilities are discovered, making it challenging for organizations to keep up with the pace. According to the National Vulnerability Database, there were over 17,000 new vulnerabilities discovered in 2020 alone (Source: NVD). This sheer volume of new vulnerabilities makes it challenging for organizations to stay on top of vulnerability management. Another limitation of vulnerability management is the lack of sufficient resources and budget. Vulnerability management requires significant resources, including personnel, tools, and budget. However, many organizations struggle to allocate the necessary resources to vulnerability management. Finally, vulnerability management is also limited by prioritization and false positives. With the sheer volume of vulnerabilities discovered daily, it can be challenging for organizations to prioritize which vulnerabilities to remediate first. Additionally, many vulnerability scanners and tools generate false positives, which can lead to wasted resources and effort [27].

6 Conclusion

In this article, we explored in a global view the evolution of cybersecurity from reliance on traditional security equipment to proactive managing vulnerabilities. We highlighted the critical importance of understanding and addressing vulnerabilities, as they represent potential entry points for cyber threats.

We examined the types of vulnerabilities that organizations face, the steps necessary for effective management, and the essential tools and frameworks that facilitate this process. We also presented some case studies to show how various big organizations need to implement vulnerability management strategies to reinforce the need for continuous monitoring and risk assessment. Furthermore, some successful stories of organizations that use vulnerability management programs to empower the security of their systems. And at the end, we try to mention some limitations of implementing vulnerability management frameworks and tools which can be a challenge for me to write my next article.

As a call to action, organizations must prioritize the development and implementation of robust vulnerability management programs. This includes investing in the necessary tools, training personnel, and fostering a culture of cybersecurity awareness. By doing so, organizations can not only protect their assets but also ensure the continuity of their operations in an increasingly complex cyber landscape.

References

1. A. Blali, S. Dargaoui, M. Azrour, A. Guezzaz, F. Amounas, and A. Alabdulatif, "Analysis of deep learning-based intrusion detection systems in IoT environments," *EDPACS*, pp. 1–35, 2025.

2. S. Dargaoui, M. Azrour, A. El, A. Guezzaz, A. Alabdulatif, and S. Ahmad, "ECC-Based Anonymous and Multi-factor Authentication Scheme for IoT Environment," vol. 21, no. 1, 2025.
3. K. Bella, M. Mohy-Eddine, A. Guezzaz, S. Benkirane, and M. Azrour, "Efficient IoT intrusion detection with XGBoost and feature selection," in *Recent Advances in Internet of Things Security*, CRC Press, 2025, pp. 82–92.
4. S. Dargaoui *et al.*, "IoT-Driven Smart Agriculture: Security Issues and Authentication Schemes Classification," in *Proceeding of the International Conference on Connected Objects and Artificial Intelligence (COCIA2024)*, Y. Mejdoub and A. Elamri, Eds., Cham: Springer Nature Switzerland, 2024, pp. 61–66. https://doi.org/10.1007/978-3-031-70411-6_10.
5. T. S. Mohamed, "Integrated A Robust Intelligent System For Secure Network," *Babylonian Journal of Internet of Things*, vol. 2025, pp. 66–76, 2025.
6. S. S. Batool, W. A. Adil, R. A. Talani, R. Raja, S. Abbas, and S. M. S. Bukhari, "LEVERAGING AI TO IDENTIFY ANOMALIES IN ELECTRICAL SYSTEMS AND COMMUNICATION NETWORKS, SAFEGUARDING CRITICAL INFRASTRUCTURE AGAINST CYBER-ATTACKS," *Spectrum of Engineering Sciences*, vol. 3, no. 3, pp. 452–472, 2025.
7. S. Dargaoui, M. Azrour, A. El Allaoui, A. Guezzaz, and S. Benkirane, "Machine learning for security boosting in Internet of Things environments," in *Recent Advances in Internet of Things Security*, CRC Press, 2025, pp. 63–71.
8. H. Assmi *et al.*, "A Robust Security Detection Strategy for Next Generation IoT Networks," *CMC*, vol. 82, no. 1, pp. 443–466, 2025, https://doi.org/10.32604/cmc.2024.059047.
9. S. Amaouche, A. Guezzaz, S. Benkirane, and M. Azrour, "A comparative analysis of random forest and isolation forest intrusion detection systems," in *Recent Advances in Internet of Things Security*, CRC Press, 2025, pp. 181–193.
10. M. Mohy-eddine, A. Guezzaz, S. Benkirane, and M. Azrour, "Malicious detection model with artificial neural network in IoT-based smart farming security," *Cluster Computing*, pp. 1–16, 2024.
11. C. Hazman, A. Guezzaz, S. Benkirane, and M. Azrour, "Enhanced IDS with Deep Learning for IoT-Based Smart Cities Security," *Tsinghua Sci. Technol.*, vol. 29, no. 4, pp. 929–947, Aug. 2024, https://doi.org/10.26599/TST.2023.9010033.
12. S. Khan *et al.*, "Manufacturing industry based on dynamic soft sensors in integrated with feature representation and classification using fuzzy logic and deep learning architecture," *Int J Adv Manuf Technol*, Jun. 2023, https://doi.org/10.1007/s00170-023-11602-y.
13. N. Meenakshi *et al.*, "Efficient Communication in Wireless Sensor Networks Using Optimized Energy Efficient Engroove Leach Clustering Protocol," *Tsinghua Sci. Technol.*, vol. 29, no. 4, pp. 985–1001, Aug. 2024, https://doi.org/10.26599/TST.2023.9010056.
14. P. Wang, Z. Zhu, Q. Chen, and W. Dai, "Text Reasoning Chain Extraction for Multi-Hop Question Answering," *Tsinghua Sci. Technol.*, vol. 29, no. 4, pp. 959–970, Aug. 2024, https://doi.org/10.26599/TST.2023.9010060.
15. R. Liggett, J. R. Lee, A. L. Roddy, and M. A. Wallin, "The dark web as a platform for crime: An exploration of illicit drug, firearm, CSAM, and cybercrime markets," *The Palgrave handbook of international cybercrime and cyberdeviance*, pp. 91–116, 2020.
16. A. S. George and A. H. George, "The emergence of cybersecurity medicine: protecting implanted devices from cyber threats," *Partners Universal Innovative Research Publication*, vol. 1, no. 2, pp. 93–111, 2023.
17. A. U. Patel *et al.*, "Cybersecurity and information assurance for the clinical laboratory," *The journal of applied laboratory medicine*, vol. 8, no. 1, pp. 145–161, 2023.
18. T. W. House, *Fact sheet: Biden administration and private sector leaders announce ambitious initiatives to bolster the nation's cybersecurity*. August, 2021.
19. A. S. George, T. Baskar, and P. B. Srikaanth, "Cyber threats to critical infrastructure: assessing vulnerabilities across key sectors," *Partners Universal International Innovation Journal*, vol. 2, no. 1, pp. 51–75, 2024.
20. H. Balisane, E. I. Egho-Promise, E. Lyada, and F. Aina, "Towards Improved Threat Mitigation In Digital Environments: A Comprehensive Framework For Cybersecurity Enhancement," *International Journal of Research-GRANTHAALAYAH*, vol. 12, no. 5, 2024.

21. M. Al Kinoon, "A Comprehensive and Comparative Examination of Healthcare Data Breaches: Assessing Security, Privacy, and Performance," 2024.
22. T. Sterner *et al.*, "Quick fixes for the environment: part of the solution or part of the problem?," *Environment: Science and Policy for Sustainable Development*, vol. 48, no. 10, pp. 20–27, 2006.
23. S. Deb, "Ticketmaster Confirms Data Breach. Here's What to Know.," *The New York Times*, May 31, 2024. Accessed: May 28, 2025. [Online]. Available: https://www.nytimes.com/2024/05/31/business/ticketmaster-hack-data-breach.html.
24. K. Stoddart, *Cyberwarfare: threats to critical infrastructure.* Springer Nature, 2022.
25. A. S. George, "When trust fails: Examining systemic risk in the digital economy from the 2024 crowdstrike outage," *Partners Universal Multidisciplinary Research Journal*, vol. 1, no. 2, pp. 134–152, 2024.
26. P. Banerjee, "Crowdstrike cyber incident vs. past major cyber incidents: Analysis and solutions," 2024.
27. "A Survey on RFID Mutual Authentication Protocols Based ECC for Resource-Constrained in IoT Environment | SpringerLink." Accessed: May 27, 2025. [Online]. Available: https://link.springer.com/chapter/https://doi.org/10.1007/978-3-031-48573-2_28.

Evaluating Deep Learning Approaches for Intrusion Detection in IoT Networks

Abdeslem Blali, Souhayla Dargaoui, Mourade Azrour, Azidine Guezzaz, and Abdulatif Alabdulatif

Abstract This paper examines the security challenges posed by the rapid growth of the Internet of Things in sectors such as industrial, health, and agriculture. As IoT systems become more widespread, cyberattacks increasingly threatened them. To address these security issues, we conducted a review of 10 papers on different intrusion detection systems for the Internet of Things. The goal was to see how well they work. In addition, to discover ways of improvement. We show that deep learning-based intrusion detection systems could improve the way we detect online attacks. Especially when using neural networks, these systems can better detect and respond to malicious activity. The combination of machine learning and intrusion detection systems appears to have the potential to improve the security of Internet of Things networks, providing stronger protection against cyberattacks. This is where convolutional neural networks come in. Especially when tested with different datasets, this deep learning model showed very high accuracy in protecting Internet of Things networks. This highlights the importance of using sophisticated algorithms to meet the growing challenges of cyber threats against IoT systems.

Keywords IDS · IoT network security · Deep learning · CNN · LSTM · Neural networks

---

A. Blali (✉) · S. Dargaoui · M. Azrour
IMIA Laboratory, MSIA Team, Faculty of Sciences and Techniques, Moulay Ismail University of Meknes, Errachidia, Morocco
e-mail: ab.blali@edu.umi.ac.ma

A. Guezzaz
Higher School of Technology Essaouira, Cadi Ayyad University, Marrakech, Morocco
e-mail: a.guezzaz@uca.ma

A. Alabdulatif
Department of Computer Science, College of Computer, Qassim University, Buraydah, Saudi Arabia
e-mail: ab.alabdulatif@qu.edu.sa

M. Azrour and A. Alabdulatif (eds.), *AI-Driven Security for Next-Generation IoT Systems*, https://doi.org/10.1007/978-3-032-08784-3_2

1 Introduction

The Internet of Things (IoT) will connect over 30 billion devices by 2025. This makes it weak for attacks [1]. Network security is critical to stopping dangers. These dangers are denial-of-service attacks, stealing information, and changing or deleting data. Today, more and more devices are connected. Network security must ensure that information stays accessible, protected, and complete. In addition to firewalls, intrusion detection systems (IDSs) are critical to protect against various risks [2].

Over the past decade, to improve the ability of IDSs to detect malicious activity, researchers have been investigating machine learning and deep learning (DL) algorithms. Machine learning can automatically extract useful information from massive datasets. Machine learning-based IDSs are more successful at detecting attacks when sufficient training data is available. What's more, these systems are relatively easy to build and operate because you do not need to know much about them [3–7].

DL, a type of machine learning, uses models engineered to function like the neural networks of the human brain. These models understand data and apply it across different domains, solving issues associated with IoT devices. It uses multiple layers of artificial neural networks to understand data on multiple scales.

When using big data, DL works better than normal machine learning. But, because of more network traffic and security danger, IDSs still have a problem finding attack-fast [8].

This paper talks about how these systems can make IoT devices more safe. It looks at new research about IDS with DL. It will also look at the limitations of each of these systems. These include what DL algorithms will be used in 2022–2024, which DL algorithms will be selected, how complex each suggestion will be, and what types of attacks will be included in datasets used to develop IDSs based on DL.

The remainder of the paper is structured as follows. Section 2 outlines the main security challenges in IoT environments. Section 3 discusses the computational and energy constraints limiting the deployment of DL models on IoT devices. Section 4 reviews recent related works on deep learning-based intrusion detection in IoT. The selection criteria for the studies included in this review are described in Sect. 5. The datasets most frequently used to train and evaluate intrusion detection models in the IoT are shown in Sect. 6. Typical evaluation metrics are used to compare various models in Sect. 7. The most important results and limitations are discussed in Sect. 8. The paper is concluded and future research directions are suggested in Sect. 9.

2 IoT Security Challenges

The IoT has revolutionized the technology! It makes easier the connection and the exchange of information between devices, regardless of type. From simple switches to sophisticated home automation, IoT gives all devices a voice and helps automate, monitor, and control them. But this progress also entails significant security threats.

The greatest challenge is the heterogeneity of IoT devices. Some are small and straightforward, like temperature and motion sensors. Some are complicated like smart TVs, security cameras, or voice assistants. These differences make it very difficult to apply a single, unified set of security rules or protocols across all devices [9–11].

Most of the IoT hardware platforms can only offer small computational power, memory capacity and software protection. As such, they are easy to attack and difficult to secure using the conventional approaches. Amit adds that firmware updater releases are sometimes spotty or nonexistent, causing devices to remain vulnerable over time. Some devices may never get updates regardless. A lax update policy also introduces security holes throughout the network.

User privacy is another growing concern. IoT deployments gather a huge amount of personal or private data. Without adequate data protection measures or policy, information can become exposed, abused, or stolen. Weak or lacking authentication methods add insult to injury, making it easy for attackers to infiltrate devices and systems without proper validation. Other systems still rely on default passwords or send data in the clear [1, 3].

This work provides a study of the main security challenges for IoT systems. It will be centered around device fragmentation, no standard mechanism for updates, which causes privacy risks, and secure and scalable authentication methods. It is important to have an understanding of these challenges for the design and deployment of effective and efficient solutions. These weaknesses must be overcome if IoT networks are to be secure, robust, and trusted in operational settings [12] (Fig. 1).

Due to their generally low processing power and lack of robust built-in security features, IoT systems' diverse range of rooms, device types, and use cases expands the attack surface [13]. They are more susceptible to attacks since they have weak hardware and little software protection due to their low energy consumption. The diversity of devices, such as IP cameras, smart thermostats, health monitors, home assistants, and others, makes it difficult to implement consistent security. Such a broad ecosystem cannot be effectively protected by a single solution, and making it secure and adaptable would require highly detailed work that is time-consuming and hard to maintain.

One of the primary flaws, though, stems from overlooked or postponed firmware updates. The software of many IoT systems is old and never gets updated, which causes them to be a sitting duck for attackers. Often, the manufacturers of the device

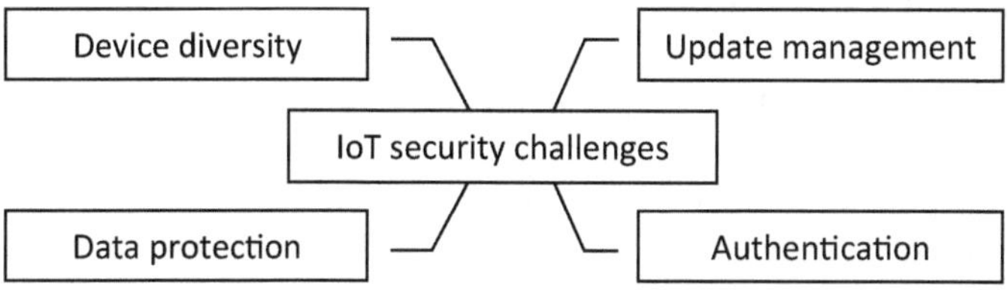

Fig. 1 Key IoT security challenges

don't have the resources to push an update or distribute a security patch. Users often also are not informed when upgrades are ready. Unattended and unwatched, these devices run out-of-date versions indefinitely. This negligence results in higher vulnerability to damages such as malware infection, sensitive information leakage and unauthorized remote operation.

The other big issue is that there is not good privacy protection. And a significant amount of the IoT devices capture and save sensitive information such as user behaviors, health records, location information, audio or video feeds, When this data is sent over unsecure or poorly-encrypted networks, it can be easily intercepted and exploited by hackers. In other cases, no encryption is used on the device at all, allowing the data to be downloaded and accessed freely. This is a significant issue of security, privacy and unauthorized access. When breached, personal information might be stolen, altered, or sold without the user's consent.

Authentication still an uphill task in IoT design. As the number of connected devices grows, the ability to verify each device and user is challenged. Most IoT environments fall short of scaleable authentication mechanism. Fundamental security practices like the use of unique IDs or two-factor authentication are all too often absent. Attackers can instead inject bogus devices or masquerade as users when there are no mechanisms to validate the identity of each element. This introduces high risk, particularly where sensitive environments such as smart grids, industrial systems or healthcare platforms are concerned. Inadequate authentication can give remote attackers the ability to take over vital infrastructure, disable necessary services, and manipulate devices. Stronger, scalable authentication techniques designed for IoT systems, standardized update policies, and improved encryption are all necessary to meet these challenges.

3 Resource Constraints in IoT

Because IoT devices have limited hardware and energy capabilities, intrusion detection in IoT environments presents special challenges. The efficiency, deployment, and design of DL-IDS are all directly impacted by these limitations.

3.1 Limited Computing Power

Low-power microcontrollers or processors, like the ARM Cortex-M series, which have limited CPU frequencies and no hardware support for floating-point operations, power the majority of IoT devices. Due to limited processing power, these hardware limitations make it challenging devices to carry out real-time DL inference. Furthermore, using these devices directly makes it impractical to implement complex neural network architectures like recurrent neural networks (RNNs) or convolutional neural networks (CNNs). The limited computational resources also limit the frequency of model updates or retraining, which makes it harder to adjust to changing threats in dynamic environments. For instance, a Raspberry Pi 3B+ can only effectively run small models. System crashes or noticeable delays occur when a full CNN is run on the device.

3.2 Memory Constraints

IoT devices usually have limited storage capacities between 512 KB and 1 GB and very little RAM, between 32 KB and 256 MB. The size and number of layers that deployable DL models can have are constrained by these memory constraints. Additionally, they restrict the use of high-dimensional input features, which are frequently necessary for precise intrusion detection. It also becomes impractical to store historical data locally for time-based analysis or training. Models must therefore be compressed using optimization methods like pruning or quantization in order to lower memory consumption without sacrificing usable performance.

3.3 Energy Limitations

Many IoT nodes have stringent power consumption limits because they run on batteries or employ energy-harvesting technologies. DL algorithms use a lot of energy, which these devices frequently cannot handle, especially during inference or training. Battery life is directly shortened by high CPU usage, and frequent data transfers to cloud server's further increase power consumption. Furthermore, it can be challenging to maintain real-time monitoring while using the low-power modes that are common for IoT devices. Energy-efficient DL models and edge-assisted architectures that offload computation from devices with limited resources are required to overcome these limitations.

3.4 Communication Overhead

In IoT environments, data transmission entails high latency and energy costs. Low-bandwidth or unreliable communication protocols like LoRa, ZigBee, or MQTT are commonly used by IoT devices; these protocols are not intended for high-volume or real-time data exchange. Because raw data requires a lot of energy and bandwidth, uploading it to the cloud is inefficient. Under these limitations, centralized training methods are unfeasible due to the unsustainable nature of continuous data transfer. Performance deterioration also affects applications that need low latency. IoT devices should use lightweight local pre-processing to reduce the amount of data transmitted over the network in order to address these problems.

3.5 *Hardware Heterogeneity*

High heterogeneity, including variations in device types, operating systems, and sensor capabilities, is a defining feature of IoT systems. Due to differences in software environments and hardware specifications, a DL model that has been developed and optimized for one device might not work with another. It can be difficult to maintain consistency in performance and behavior across various devices, frequently necessitating hardware-specific configuration and tuning. Deep learning-based IDSs must therefore be made to be both portable and flexible. To guarantee greater compatibility and simpler deployment, middleware solutions that abstract hardware differences can be used instead [14].

3.6 *Implications for Deep Learning-Based IDS*

Because of the inherent resource constraints in IoT environments, DL models can only be directly deployed on devices with very small architectures, such as those developed in the Tiny ML domain. For more complex tasks, computation must be offloaded to edge devices, which can handle higher processing demands while remaining closer to the data source than cloud solutions. These limitations force developers to carefully balance model accuracy with resource usage. Finding this balance requires the use of optimization techniques like quantization, pruning, and knowledge distillation, which can reduce model size and computational load without appreciably affecting performance [15].

4 Related Works

The explosive growth of IoT, which is transforming many areas of modern technology, has introduced significant security challenges that require the deployment of robust IDS. According to the discussion in references, DL is a viable and effective way to deal with these issues in the context of the IoT. When it comes to protecting IoT systems, DL has a number of benefits. Large data sets are processed, intricate and nonlinear patterns are found, detection accuracy is increased, and persistent problems like high false alarm rates, where traditional metrics fall short, are addressed. These features allow DL to perform better than conventional IDS techniques, which frequently have trouble handling dynamic and high-dimensional IoT data [16].

The research shown in [13, 17–19] emphasizes how advanced AI techniques, particularly DL and ML, can be used to improve IoT security systems. These studies stress the significance of using diverse models, datasets, and assessment frameworks to identify different types of intrusions and anomalies. Despite being widely used because of its simplicity and reduced computational requirements, machine learning often performs poorly in the face of unknown or zero-day attacks. On the other hand, DL not only increases threat detection accuracy but also lessens false positives and more skillfully manages class imbalance in datasets.

DL for IoT security still has a number of drawbacks despite its potential. The significant financial and computational expenses involved in creating and maintaining DL-based IDS, along with worries about user privacy and data security, are some of the main obstacles. For DL-based solutions to be widely adopted and effective in actual IoT environments, these challenges must be addressed. Performance, cost, and privacy must be balanced as research progresses in order to make DL a dependable and expandable tool for IoT security.

5 Criteria for Studie Selection

The reviewed studies are chosen based on predetermined standards intended to guarantee quality and relevance in the field. Only recent studies that used DL models, such as Deep Neural Networks (DNNs), CNN, and LSTM, or Gated Recurrent Units (GRUs), were taken into consideration in order to maintain a high standard. These models were chosen because of their demonstrated ability to manage intricate data sequences and patterns that are frequently encountered in IoT security scenarios.

Research using the most recent datasets and reporting the highest-performing evaluation metrics (e.g., accuracy, precision, recall, or F_1-score) were prioritized. This emphasis made it easier to weed out antiquated methods and guaranteed that the results represented the most recent developments in technology and trends. We mostly used Google Scholar to refine the scope and increase the accuracy of the literature selection procedure. The most pertinent, influential, and peer-reviewed scholarly works about the application of DL to IoT security were found, accessed, and assessed using this platform.

6 Dataset Description

6.1 Overview

The public datasets used in the studies examined in this review are contrasted in this section. Deep learning-based IDSs in IoT environments were trained and assessed using these datasets. Their distribution of classes, balance, types of attacks, year of release, and applicability to IoT scenarios vary. The main features of each dataset are summed up in a comparative Table 1.

Table 1 Comparative table

Dataset	Year	Attack types	No. of classes	IoT-specific	Balanced
CIC-IDS2017	2017	DDoS, Infiltration, Brute Force	15	No	No
CSE-CIC-IDS2018	2018	Botnet, DoS, Brute Force	15	Partially	No
NSL-KDD	2009	Probe, R2L, U2R, DoS	5	No	No
UNSW-NB15	2015	10 types (Fuzzers, Backdoor, etc.)	10	No	No
N-BaIoT	2018	Botnet (Mirai, Gafgyt)	2	Yes	Yes
MQTT-IoT-IDS2020	2020	MQTT-specific attacks	5	Yes	Partial
Bot-IoT	2018	DDoS, Reconnaissance, etc.	5+	Yes	No
ToN_IoT	2020	Malware, Ransomware, etc.	10+	Yes	No
IoT-23	2020	Botnet, Port scanning, etc.	Varies	Yes	Partial
CICDIoT2023	2023	Zero-day, DoS, Botnet, etc.	10+	Yes	No
Edge-IoT	2022	Edge-Focused Attacks (DDoS, etc.)	Varies	Yes	Partial
NDL-KDD	2022	U2R, R2L, DoS, Probe	5	No	Yes

6.2 Dataset Characteristics

IoT Relevance: The only datasets created especially for IoT-based traffic are N-BaIoT, ToN_IoT, Bot-IoT, and CICDIoT2023. Others, like UNSW-NB15 and NSL-KDD, are more general and might miss behavior unique to the IoT.

Attack Diversity: The threats covered by different datasets vary. A wide range of recent attacks (such as ransomware, spoofing, and zero-day attacks) are included in CICDIoT2023, ToN_IoT, and Edge-IoT. On the other hand, more conventional attack types are included in NSL-KDD and NDL-KDD.

Number of Classes: the number of output classes influences the granularity of detection. Model evaluation under complex threat conditions is made possible by multi-class datasets (such as CSE-CIC-IDS2018 and CICDIoT2023). N-BaIoT and other binary datasets make detection easier but restrict the depth of analysis.

Balance: By reducing bias during training, N-BaIoT and NDL-KDD are comparatively balanced. The majority are extremely unbalanced, necessitating the use of class rebalancing strategies like weighted loss or oversampling.

Recency: CICDIoT2023, Edge-IoT, and MQTT-IoT-IDS2020 are appropriate for contemporary IDS research since they represent new devices, protocols, and threat vectors.

7 Performance Analysis of the Different Models Studied

Maintaining security in such settings is becoming more and more difficult as the IoT continues to grow. Intruders can be detected and stopped using a variety of algorithms and techniques, each of which has advantages and disadvantages depending on the circumstances. In addition to providing a survey of the body of research on the various network and host architectures, this section illustrates models like CNN architectures, LSTM networks, and hybrid models and how they can improve IoT security [20].

The DL model functions well and there are numerous crucial aspects of the machine to examine. Datasets contain information that can be used to test and teach algorithms about how data will be used. Accuracy indicates the model's quality, but when data is unequal, it may be inaccurate. Recall checks whether the model finds all good examples, and precision checks whether the positive response is accurate. When mistakes in positive and negative classifications have serious repercussions, the F_1-score is essential.

A CNN model trained on the CSECICIDS2018 data set was critically examined in the experimental setup by paper [21], demonstrating an accuracy of 99.65%. For intrusion detection tasks, it shows good performance with precision, recall, and F_1 of 99.16%, 98.70%, and 99.09%, respectively.

Based on the MQTT-IoT IDS2020 data set, the hybrid CNN-LSTM model presented in the paper [22] demonstrated good capability in IoT security detection with accuracy, precision, recall, and F_1 of 99.86%, 99.80%, 99.87%, and 99.83%, respectively.

On the N-BaIoT training set, the DNN model detailed in the paper [23] achieved an accuracy of 97.21%. With an F_1-score of 88.48%, a precision of 91.41%, and a recall of 87.31%, this model did reasonably well but fell just short of CNNs.

On the CICDIoT2023 data set, different models were compared in a paper [24]. The CNN model achieved the highest classification accuracy, with a rate of 96.37%, and an equivalent precision and recall of 96.37%, along with an accompanying F_1-score of 95.51%. With an F_1-score of 88.51% and classification accuracy of 88.64%, precision of 91.20%, and recall of 88.64%, JNN did not fare as well as the other models. However, the RNN model outperformed the other models, achieving an accuracy of 96.52%, an F^1-score of 95.73%, and estimates for precision and recall of 96.25%.

An LSTM model is tested on three distinct datasets in Paper [25]. Its results on NSL-KDD include 81.10% accuracy, 92.10% precision, 73.20% recall, and an F_1-score of 81.50%. The accuracy increases to 86.60% on UNSW-NB15, with an

F_1-score of 89.10%, 81.10% precision, and 98.80% recall. It achieves 87.30% accuracy, 78.40% precision, 88.00% recall, and an F_1-score of 82.90% on ToN_IoT, demonstrating variable performance depending on dataset properties.

Several models are compared on the Bot-IoT dataset in paper [26]. With an F_1-score of 99.80%, the CNN model attains 99.70% accuracy, 99.60% precision, and 99.90% recall. With an F_1-score of 99.80%, 99.80% accuracy, 99.70% precision, and 100.00% recall, the LSTM model performs marginally better. There aren't many differences between these architectures, as the GRU model achieves 99.60% accuracy, 99.60% precision, 100.00% recall, and an F_1-score of 99.80%.

Three models are examined using the CIC-IDS2017 dataset in paper [27]. With an F_1-score of 84.60%, 80.85% precision, 84.60% recall, and 94.61% accuracy, the DNN model is the least accurate. With 97.67% accuracy, 94.96% precision, 95.00% recall, and an F_1-score of 93.55%, the LSTM model performs better than it. With 98.61% accuracy, 97.05% precision, 96.95% recall, and an F_1-score of 98.09%, the CNN model performs the best, demonstrating its efficacy in network intrusion detection.

The CNN-based model performs exceptionally well on all metrics when evaluated on the IoT-23 data set, according to study [28]. The model achieves 99.90% accuracy, indicating a highly effective classification. With the least amount of false positives and false negatives, the model can distinguish between malicious and legitimate traffic. The CNN approach is highly appropriate for IoT security since the F_1-value of 99.90% validates the positive correlation between precision and recall.

In [29], an LSTM-GRU model is tested on both the IoT-23 dataset and the UNSW-NB15 dataset. The model's accuracy is 98.12%, with precision, recall and F_1-scores all above 98%, demonstrating how robust the model is. When applied to UNSW-NB15, the model performs almost perfectly, achieving an accuracy of 99.98% and an F_1-score of 99.99%. These results demonstrate the adaptability of the LSTM-GRU model across multiple data sets and demonstrate its effectiveness in handling complex network traffic scenarios in IoT settings.

The study [30] assesses the performance of a model based on LSTM on three data sets: Bot-IoT, Edge-IIoT, and NSL-KDD. On Bot-IoT, the model demonstrates its strong ability to detect network intrusions, achieving 99.94% accuracy. On the Edge-IIoT data set, the model achieves a near-perfect accuracy of 99.99%, with 100% precision and recall, demonstrating exceptional reliability in industrial IoT environments. On the NSL-KDD dataset, confirming its robustness in different network traffic scenarios, the model maintains its high effectiveness with 99.81% accuracy, 99.71% precision, and 99.87% recall. These findings demonstrate that LSTMs are adaptable and effective for intrusion detection in different IoT and industrial networks (Table 2).

Table 2 Performance of the different models studied

Paper	Algorithm	Dataset	Accuracy (%)	Precision (%)	Recall	F_1-score (%)
[21]	CNN	CSE-CIC-IDS2018	99.65	99.16	98.70	99.09
[22]	CNN-LSTM	MQTT-IoT-IDS2020	99.86	99.80	99.87	99.83
[23]	DNN	N-BaIoT	97.21	91.41	87.31	88.48
[24]	CNN	CICDIoT2023	96.37	96.15	96.37	95.51
	DNN		88.64	91.20	88.64	88.51
	RNN		96.52	96.25	96.52	95.73
[25]	LSTM	NSL-KDD	81.10	92.10	73.20	81.50
		UNSW-NB15	86.60	81.10	98.80	89.10
		ToN_IoT	87.30	78.40	88.00	82.90
[26]	CNN	Bot-IoT	99.70	99.60	99.90	99.80
	LSTM		99.80	99.70	100.00	99.80
	GRU		99.60	99.60	100.00	99.80
[27]	DNN	CIC-IDS2017	94.61	80.85	84.60	84.60
	LSTM		97.67	94.96	95.00	93.55
	CNN		98.61	97.05	96.95	98.09
[28]	CNN	IoT-23	99.90	99.83	99.97	99.90
[29]	LSTM-GRU	IoT-23	98.12	98.06	98.31	98.18
		UNSW-NB15	99.98	99.99	99.98	99.99
[30]	LSTM	bot-iot	99.94	99.94	99.94	–
		Edge-Iiot	99.99	100	100	–
		NDL-KDD	99.81	99.71	99.87	–

8 Discussion

The thorough assessment of several DL models on a variety of datasets highlights how effective each method is at intrusion detection for IoT environments. A more thorough grasp of each model's performance under various circumstances and use cases is made possible by this comparison. In particular, CNNs are a reliable choice for analyzing IoT network traffic because they continuously exhibit high performance levels across a range of benchmarks. CNN, for instance, obtained a 99.65% accuracy on the CSE-CIC-IDS2018 dataset in [12], and an even greater 99.90% accuracy on the IoT-23 dataset in [22]. The ability of CNN models to extract significant features from both structured and unstructured network data is validated by these robust and consistent results.

In a similar vein, hybrid models that combine CNN and LSTM networks have demonstrated remarkable performance. These architectures are made to take advantage of the LSTM's prowess in identifying temporal patterns and the CNN's capacity to identify spatial features. By combining several feature extraction techniques; this hybrid approach improves IDSs, as illustrated in [13], where a CNN-LSTM model obtained 99.86% accuracy on the MQTT-IoT-IDS2020 dataset. In dynamic network environments, the combination of CNN and LSTM provides enhanced robustness.

DNNs, on the other hand, perform worse than CNN or LSTM-based architectures, despite still being useful in many use cases. For example, DNN had a recall of only 87.31% in [17] despite achieving 97.21% accuracy on the N-BaIoT dataset, suggesting a greater number of undetected attacks. On the CICDIoT2023 dataset, the DNN achieved an accuracy of 88.64% in [18], which is significantly less than CNN or RNN results on the same benchmark. These results imply that DNNs may not have the specific capability required to capture the hierarchical or sequential structure of IoT traffic.

Particularly when handling time-series or sequential data, which are frequently found in IoT environments, LSTM-based models continue to demonstrate excellent performance. With an astounding 99.80% accuracy rate, LSTM beat both CNN and GRU on the Bot-IoT dataset in [20]. Additionally, results showing an LSTM-GRU hybrid achieving 99.98% accuracy on the UNSW-NB15 dataset are presented in [23]. When identifying changing threats or patterns of behavior over time, this performance level facilitates the use of recurrent architectures.

The performance of the model is also significantly influenced by the data itself. On the CIC-IDS2017 dataset, CNN, DNN, and LSTM achieved 98.61%, 94.61%, and 97.67% accuracy, respectively, as demonstrated in [21], illustrating how various models react to various data features. On datasets as Edge-IIoT and IoT-23, LSTM models almost achieved perfect detection in [24], confirming their applicability for real-world deployments. These variances highlight the necessity of adaptable, dataset-specific models that can accommodate various traffic types, scales, and structures.

Although conventional performance metrics such as accuracy, precision, recall, and F_1-score are crucial for assessing detection capabilities, they are not the sole determinants of a model's usefulness in IoT environments. IoT devices frequently have very strict energy, memory, and processing power limits. In real-time or on-device applications, a highly accurate model that requires a lot of processing power might not be practical. Performance must be balanced with limitations such as inference speed, memory usage, and energy efficiency for practical deployment. To guarantee that DL models can be effectively implemented in operational IoT networks, their scalability, hardware compatibility, and operational cost must all be taken into account during the evaluation process.

9 Conclusion

In conclusion, this study unequivocally demonstrates that IDSs, or IDSs, are essential to protecting IoT networks from a variety of dangers. IDS performance has improved significantly as a result of the development of artificial intelligence (AI), particularly in recent years, which has made it possible to detect suspicious or malicious activity in these complex environments more quickly and accurately. DL algorithms have demonstrated a strong ability to manage and process massive volumes of data, which makes them ideal for real-time intrusion detection tasks as IoT network traffic continues to grow rapidly. You can gain a more thorough grasp of how to construct and optimize secure IoT infrastructures by evaluating various research studies to determine the advantages and disadvantages of each approach.

The significance of selecting appropriate software environments, testing platforms, and development tools for IoT system design is also emphasized in this paper. Python is still the most widely used programming language for creating DL models, and because of its community support, ease of use, and flexibility, well-known libraries like TensorFlow and Keras are regularly used. Because platforms like Cooja allow for better control and repeatability of experiments, most researchers opt to simulate their models rather than relying on physical IoT devices like Raspberry Pi or Arduino, which are less practical for large-scale testing. The volume, features, and focus of the datasets used in these studies vary greatly. To improve the accuracy of detection models and reflect real-world traffic, many incorporate millions of entries. Particularly well liked DL models that combine spatial and temporal pattern recognition are CNN and LSTM.

In the end, maintaining secure IoT systems necessitates constant work to balance the trade-offs between real-time operation, model complexity, and detection accuracy. For researchers and practitioners looking to implement effective, scalable, and resilient IDS solutions in a variety of IoT environments, striking the correct balance between these factors continues to be a major challenge.

References

1. Dargaoui, S., Azrour, M., El, A., Guezzaz, A., Alabdulatif, A., Ahmad, S.: ECC-Based Anonymous and Multi-factor Authentication Scheme for IoT Environment, (2025).
2. Almaraz-Rivera, J.G., Cantoral-Ceballos, J.A., Botero, J.F.: Enhancing IoT Network Security: Unveiling the Power of Self-Supervised Learning against DDoS Attacks. Sensors. 23, 8701 (2023). https://doi.org/10.3390/s23218701.
3. Assmi, H., Guezzaz, A., Benkirane, S., Azrour, M., Jabbour, S., Innab, N., Alabdulatif, A.: A Robust Security Detection Strategy for Next Generation IoT Networks. Comput. Mater. Contin. 82, 443–466 (2025). https://doi.org/10.32604/cmc.2024.059047.
4. Attou, H., Guezzaz, A., Benkirane, S., Azrour, M.: A New Secure Model for Cloud Environments Using RBFNN and AdaBoost. SN Comput. Sci. 6, 188 (2025). https://doi.org/10.1007/s42979-025-03691-1.
5. Amaouche, S., Guezzaz, A., Benkirane, S., Azrour, M.: A Robust Model for Predicting Abnormal Behavior in Vehicular Networks using AdaBoost and Chi-Square. Wirel. Pers. Commun. (2024). https://doi.org/10.1007/s11277-024-11615-0.
6. Bella, K., Guezzaz, A., Benkirane, S., Azrour, M., Fouad, Y., S. Benyeogor, M., Innab, N.: An efficient intrusion detection system for IoT security using CNN decision forest. PeerJ Comput. Sci. 10, e2290 (2024). https://doi.org/10.7717/peerj-cs.2290.
7. Mohy-eddine, M., Guezzaz, A., Benkirane, S., Azrour, M.: Malicious detection model with artificial neural network in IoT-based smart farming security. Clust. Comput. 1–16 (2024).

8. Elsaid, S.A., Shehab, E., Mattar, A.M., Azar, A.T., Hameed, I.A.: Hybrid intrusion detection models based on GWO optimized deep learning. Discov. Appl. Sci. 6, 531 (2024). https://doi.org/10.1007/s42452-024-06209-1.
9. Hazman, C., Guezzaz, A., Benkirane, S., Azrour, M.: A smart model integrating LSTM and XGBoost for improving IoT-enabled smart cities security. Clust. Comput. 28, 70 (2024). https://doi.org/10.1007/s10586-024-04780-1.
10. Dargaoui, S., Azrour, M., El Allaoui, A., Guezzaz, A., Benkirane, S., Alabdulatif, A., Amounas, F.: Internet-of-Things-Enabled Smart Agriculture: Security Enhancement Approaches. In: 2024 4th International Conference on Innovative Research in Applied Science, Engineering and Technology (IRASET). pp. 1–5 (2024). https://doi.org/10.1109/IRASET60544.2024.10548705.
11. Anwar, T., Khan, G.A., Ashraf, Z., Ansari, Z.A., Ahmed, R., Azrour, M.: The Combination of Blockchain and the Internet of Things (IoT): Applications, Opportunities, and Challenges for Industry. In: Blockchain and Machine Learning for IoT Security. Chapman and Hall/CRC (2024).
12. Tawalbeh, L., Muheidat, F., Tawalbeh, M., Quwaider, M.: IoT Privacy and Security: Challenges and Solutions. Appl. Sci. 10, 4102 (2020). https://doi.org/10.3390/app10124102.
13. Islam, M.R., Rahman, R.: Best Practices for Facing the Security Challenges of Internet of Things Devices Focusing on Software Development Life Cycle, http://arxiv.org/abs/2402.07832, (2024). https://doi.org/10.48550/arXiv.2402.07832.
14. Dargaoui, S., Azrour, M., Allaoui, A.E., Guezzaz, A., Benkirane, S.: Machine learning for security boosting in Internet of Things environments. In: Recent Advances in Internet of Things Security. CRC Press (2025).
15. Mohy-Eddine, M., Guezzaz, A., Benkirane, S., Azrour, M., Bella, K.: A monitoring system with deep learning for IoT smart environments security. In: Recent Advances in Internet of Things Security. CRC Press (2025).
16. Blali, A., Dargaoui, S., Azrour,M., Guezzaz,A., Amounas,F., and Alabdulatif, A.: Analysis of deep learning-based intrusion detection systems in IoT environments. EDPACS. 0, 1–35. https://doi.org/10.1080/07366981.2025.2498222.
17. Raoufi, P., Hemmati, A., Rahmani, A.M.: Deep learning applications in the Internet of Things: a review, tools, and future directions. Evol. Intell. 17, 3621–3654 (2024). https://doi.org/10.1007/s12065-024-00949-0.
18. Ghaffari, A., Jelodari, N., pouralish, S., derakhshanfard, N., Arasteh, B.: Securing internet of things using machine and deep learning methods: a survey. Clust. Comput. 27, 9065–9089 (2024). https://doi.org/10.1007/s10586-024-04509-0.
19. Elhanashi, A., Dini, P., Saponara, S., Zheng, Q.: Integration of Deep Learning into the IoT: A Survey of Techniques and Challenges for Real-World Applications. Electronics. 12, 4925 (2023). https://doi.org/10.3390/electronics12244925.
20. Kilichev, D., Turimov, D., Kim, W.: Next–Generation Intrusion Detection for IoT EVCS: Integrating CNN, LSTM, and GRU Models. Mathematics. 12, 571 (2024). https://doi.org/10.3390/math12040571.
21. G, M., Maheswaravenkatesh, P.: A Lightweight Convolutional Neural Network based network intrusion detection and classification method for Social Internet of Things, https://www.researchsquare.com/article/rs-3795283/v1, (2024). https://doi.org/10.21203/rs.3.rs-3795283/v1.
22. AS-CL IDS: anomaly and signature-based CNN-LSTM intrusion detection system for Internet of Things. Int. J. Adv. Technol. Eng. Explor. 10, (2023). https://doi.org/10.19101/IJATEE.2022.10100187.
23. Wardana, A.A., Kołaczek, G., Warzyński, A., Sukarno, P.: Ensemble averaging deep neural network for botnet detection in heterogeneous Internet of Things devices. Sci. Rep. 14, 3878 (2024). https://doi.org/10.1038/s41598-024-54438-6.
24. Abbas, S., Bouazzi, I., Ojo, S., Hejaili, A.A., Sampedro, G.A., Almadhor, A., Gregus, M.: Evaluating deep learning variants for cyber-attacks detection and multi-class classification in IoT networks. PeerJ Comput. Sci. 10, e1793 (2024). https://doi.org/10.7717/peerj-cs.1793.

25. Sharma, B., Sharma, L., Lal, C., Roy, S.: Explainable artificial intelligence for intrusion detection in IoT networks: A deep learning based approach. Expert Syst. Appl. 238, 121751 (2024). https://doi.org/10.1016/j.eswa.2023.121751.
26. Banaamah, A.M., Ahmad, I.: Intrusion Detection in IoT Using Deep Learning. Sensors. 22, 8417 (2022). https://doi.org/10.3390/s22218417.
27. Jose, J., Jose, D.V.: Deep learning algorithms for intrusion detection systems in internet of things using CIC-IDS 2017 dataset. Int. J. Electr. Comput. Eng. IJECE. 13, 1134 (2023). https://doi.org/10.11591/ijece.v13i1.pp1134-1141.
28. Nobakht, M., Javidan, R., Pourebrahimi, A.: DEMD-IoT: a deep ensemble model for IoT malware detection using CNNs and network traffic. Evol. Syst. 14, 461–477 (2023). https://doi.org/10.1007/s12530-022-09471-z.
29. Sanju, P.: Enhancing intrusion detection in IoT systems: A hybrid metaheuristics-deep learning approach with ensemble of recurrent neural networks. J. Eng. Res. 11, 356–361 (2023). https://doi.org/10.1016/j.jer.2023.100122.
30. Hazman, C., Guezzaz, A., Benkirane, S., Azrour, M.: Enhanced IDS with Deep Learning for IoT-Based Smart Cities Security. Tsinghua Sci. Technol. 29, 929–947 (2024). https://doi.org/10.26599/TST.2023.9010033.

An Efficient Multi-factor Authentication Scheme Based on Elliptic Curve Cryptography for Secure IoT Communication

Amal Benqassmi, Mohamed Badiy, Fatima Amounas, Mourade Azrour, Salma Bendaoud, Hind Timouhin, and Ibtissam Essadik

Abstract With the evolution of the Internet of Things (IoT), Elliptic Curve Cryptography (ECC) appears to be a highly effective solution for securing devices with limited resources. Since authentication helps prevent unauthorized access and misuse of data exchanged between IoT devices, numerous lightweight IoT authentication protocols have been introduced by researchers. Many IoT authentication schemes that rely on two-factor authentication often fall short in achieving important security goals. To address this challenge, we introduce a robust three-factor authentication scheme based on ECC to authenticate both ends of the communication. We validate the security of our scheme through informal analysis and formal verification using the AVISPA tool, demonstrating its resilience against known attacks. A comparative analysis reveals that our protocol outperforms existing schemes by offering higher functionality and enhanced security, all while maintaining minimal computational cost.

Keywords IoT security · Multi-authentication · Elliptic curve cryptography · ElGamal algorithm · AVISPA

A. Benqassmi · F. Amounas · M. Azrour (✉) · S. Bendaoud · H. Timouhin
IMIA Laboratory, MSIA Team, Faculty of Sciences and Techniques Errachidia, Moulay Ismail University of Meknes, Meknes, Morocco
e-mail: mo.azrour@umi.ac.ma

A. Benqassmi
e-mail: a.benqassmi@edu.umi.ac.ma

F. Amounas
e-mail: f.amounas@umi.ac.ma

M. Badiy
National School of Artificial Intelligence and Digital, Berkane, Morocco
e-mail: m.badiy@edu.umi.ac.ma

I. Essadik
SETIME Laboratory, IPAI Team, Faculty of Sciences, Ibn Tofail University, Kenitra, Morocco
e-mail: ibtissam.essadik@uit.ac.ma

M. Azrour and A. Alabdulatif (eds.), *AI-Driven Security for Next-Generation IoT Systems*, https://doi.org/10.1007/978-3-032-08784-3_3

1 Introduction

The advent of the Internet of Things (IoT) has driven a rapid and exponential increase in the number of connected devices [1–3]. These devices often operate with limited computational and storage resources, making them incapable of handling traditional cryptographic protocols and susceptible to various security threats. Recently, IoT settings such as intelligent buildings, smart homes, and connected offices have swiftly become part of daily life [4–7]. A standard IoT framework comprises diverse micro-devices that gather and relay different kinds of data in real time. However, the use of wireless networks introduces significant vulnerabilities, as they are inherently open channels prone to unauthorized access attacks. Malicious actors can exploit these weaknesses by inserting, deleting, or modifying data to gain unauthorized access to sensitive user information or impersonate legitimate users [8]. To overcome these issues, authentication has been a key topic of in-depth research in the last two decades [9, 10], especially due to the increasing complexity of security threats, the growth of digital services, and the need for stronger and more reliable mechanisms to protect sensitive information. A wide range of lightweight authentication schemes has been developed in recent times [11–17]. Despite extensive efforts, the analysis reveals that most of these protocols have certain limitations. To address these issues, our primary contribution is the design of a secure mutual authentication protocol, which emerges as a fundamental requirement for establishing trusted communication. This paper provides the following research contributions:

- We suggest a new three-factor mutual authentication scheme based on ECC to ensure secure and reliable communication between IoT edge nodes and cloud platforms.
- Informal security analysis and simulation with AVISPA highlight the effectiveness of our scheme and its resilience.

The rest of the present chapter is organized as following. The Sect. 2 presents the proposed methodology. Section 3 details the results of the security property analysis of the proposed scheme, while Sect. 4 provides its evaluation using the AVISPA tool. Finally, Sect. 5 concludes the paper.

2 Proposed Method

In this section, we introduce an ECC-based multi-factor authentication system to ensure secure IoT communication frameworks. Table 1 displays the notations associated with the proposed method. Our protocol's architecture is shown in Fig. 1. Our system consists of two primary stages: the initialization phase and the authentication phase.

Table 1 Notations adopted in our method

Notation	Description
E	An elliptic curve over finite field Fp
P	Basis point on elliptic curve
s and Ps	Private and public keys of server
d and Pd	Private and public keys of device
R	Random number
PW, ID	Password and secret session identifier
K	Session key candidate
TS	Time stamp
H(), Next()	Hash functions
||	Concatenation operation
⊕	XOR operation

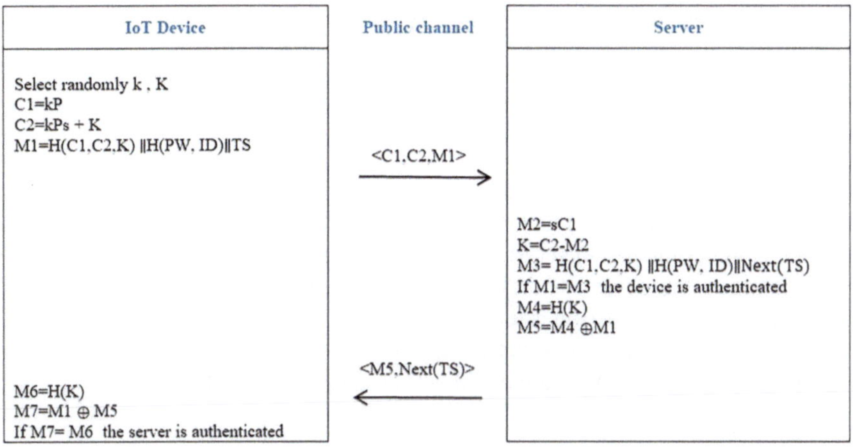

Fig. 1 Proposed authentication scheme

2.1 *Initialization Phase*

First, the server selects a random number $s \in F_p$ (finite field of prime order p) as its private key and then, it computes the corresponding public key using Eq. (1):

$$Ps = s.P \tag{1}$$

Similarly, the device selects a random number d $\in F_p$ as its private key, and then, using Eq. (2), it generates the device's public key:

$$Pd = \mathrm{d}P \tag{2}$$

2.2 Authentication Phase

This phase is carried out through the following steps:

Step 1. The device starts by generating a random value k as given in Eq. (3) and calculates a corresponding point on the elliptic curve based on Eq. (4). Next, it then randomly selects another point on the elliptic curve to calculate the encrypted message using ElGamal algorithm as illustrated in Eq. (5) and the hash value M1 as shown in Eq. (6). Finally, the device sends the message (*C1, C2, M1, TS*) to the server.

$$k \in F_p \ and \ K \in E(F_p) \tag{3}$$

$$C1 = kP \tag{4}$$

$$C2 = kPs + K \tag{5}$$

$$M1 = H(C1, C2, K)\|H(PW, ID)\|TS \tag{6}$$

Step 2. Upon receiving the message, the server computes M2 by using its private key as illustrated in Eq. (7) and the session key K as shown in Eq. (8). Then, the server calculates the parameter M3 according to Eq. (9) and compares M1 with M3 to authenticate the device. If they match, the device authentication is considered successful. If not, the session is terminated.

Next, the server calculates M4 as shown in Eq. (10) and M5 as described in Eq. (11) and then sends M5, along with a timestamp, back to the device.

$$M2 = sC1 \tag{7}$$

$$K = C2 - \mathrm{M2} \tag{8}$$

$$M3 = H((C1,\ C2,\ K)\|H(PW,\ ID)\|Next(TS)) \tag{9}$$

$$M4 = H(K) \tag{10}$$

$$M5 = M4 \oplus M1 \tag{11}$$

Step 3. Upon receiving the message, the device calculates M6 as outlined in Eq. (12) and authenticates the server by confirming whether M7 (calculated in Eq. (13)) matches M6. If the comparison is successful, mutual authentication is completed.

Otherwise, the session is terminated.

$$M6 = H(K) \tag{12}$$

$$M7 = M1 \oplus M5 \tag{13}$$

3 Security Analysis

This section outlines the security features provided by the proposed protocol, which include mutual authentication, confidentiality, and integrity. These features work together to ensure that all communications between parties are verified, kept private, and unaltered, thereby safeguarding the system against various security threats. Specifically, mutual authentication confirms the identities of both parties involved, confidentiality ensures that transmitted data remains secret from unauthorized entities, and integrity guarantees that the data has not been tampered with during transit, all contributing to a robust and secure communication framework.

- *Mutual authentication*

This service is achieved between the device and the server by checking if the reconstructed M3 matches the received M1 and by comparing M7 with M6. Consequently, our scheme successfully guarantees mutual authentication.

- *Confidentiality*

The protocol's security fundamentally relies on the Elliptic Curve Discrete Logarithm Problem (ECDLP), which ensures computational infeasibility for adversaries.

- *Integrity*

All critical messages incorporate hash-based message authentication codes: Eqs. (6) and (9) making any tampering detectable. Additionally, using Timestamp prevents replay of old messages.

4 Analysis Result in AVISPA

AVISPA represents a simple tool for the automatic assessment of various types of Internet security protocols and/or applications. Providing an expressive, flexible formal programming language for the specification of protocols as well as associated security attributes, AVISPA also incorporates a variety of back-ends which implement state-of-the-art automatic validation techniques. Based on the experimental results

OFMC
SAFE

CL-AtSe
SAFE

Fig. 2 Simulation results

achieved on an exhaustive collection of Internet security protocols, the AVISPA software package represents the latest state-of-the-art in the field of automatic security protocols. We conducted rigorous formal verification of our protocol's security properties using the AVISPA platform, a widely adopted framework for analyzing cryptographic protocols [18]. This tool evaluates protocols under two validation states: SAFE and UNSAFE. A simulation yields a SAFE result if the scheme successfully resists man-in-the-middle (MITM) attacks. Our scheme was evaluated using two AVISPA back-ends [16, 19, 20]: the OFMC and the CL-AtSe. The simulation results show that both analysis engines classify the proposed protocol as SAFE, as illustrated in Fig. 2, confirming that it successfully meets essential security objectives.

5 Conclusion

Securing IoT networks is a major challenge due to limited device resources and increasing attack risks. While many authentication schemes have been suggested, few effectively balance security and efficiency. This paper introduces a robust multi-authentication protocol based on ECC, offering strong protection against common threats. This protocol is constructed using a one-way hash function, EC-ElGamal Algorithm, and random values. Simulation results validate its robustness. Future work will explore real-world deployment, especially in sectors like healthcare, smart cities, and industrial IoT, focusing on scalability and practical adaptability.

References

1. S. Dargaoui, M. Azrour, A. El, A. Guezzaz, A. Alabdulatif, and S. Ahmad, "ECC-Based Anonymous and Multi-factor Authentication Scheme for IoT Environment," vol. 21, no. 1, 2025.
2. S. Dargaoui, M. Azrour, A. El Allaoui, A. Guezzaz, and S. Benkirane, "Machine learning for security boosting in Internet of Things environments," in *Recent Advances in Internet of Things Security*, CRC Press, 2025, pp. 63–71.
3. C. Hazman, A. Guezzaz, S. Benkirane, and M. Azrour, "Enhanced IDS with Deep Learning for IoT-Based Smart Cities Security," *Tsinghua Sci. Technol.*, vol. 29, no. 4, pp. 929–947, Aug. 2024, https://doi.org/10.26599/TST.2023.9010033.

4. S. Dargaoui *et al.*, "IoT-Driven Smart Agriculture: Security Issues and Authentication Schemes Classification," in *Proceeding of the International Conference on Connected Objects and Artificial Intelligence (COCIA2024)*, Y. Mejdoub and A. Elamri, Eds., Cham: Springer Nature Switzerland, 2024, pp. 61–66. https://doi.org/10.1007/978-3-031-70411-6_10.
5. S. Dargaoui *et al.*, "Security Issues in Internet of Medical Things," in *Blockchain and Machine Learning for IoT Security*, Chapman and Hall/CRC, 2024, pp. 77–91.
6. M. Mohy-Eddine, M. Azrour, J. Mabrouki, F. Amounas, A. Guezzaz, and S. Benkirane, "Embedded Web Server Implementation for Real-Time Water Monitoring," in *Advanced Technology for Smart Environment and Energy*, J. Mabrouki, A. Mourade, A. Irshad, and S. A. Chaudhry, Eds., in Environmental Science and Engineering., Cham: Springer International Publishing, 2023, pp. 301–311. https://doi.org/10.1007/978-3-031-25662-2_24.
7. M. Saqib and A. H. Moon, "A systematic security assessment and review of internet of things in the context of authentication," *Computers & Security*, vol. 125, p. 103053, 2023.
8. H. Timouhin, F. Amounas, and M. Badiy, "enhanced IoT security," *Recent Advances in Internet of Things Security*, p. 39, 2025.
9. R. R. Sonth, Y. R. Pranamya, N. Harish Kumar, and G. Deepak, "A Survey on Methodologies and Algorithms for Mutual Authentication in IoT Devices," in *Evolutionary Computing and Mobile Sustainable Networks: Proceedings of ICECMSN 2020*, Springer, 2021, pp. 309–315.
10. B. Hammi, A. Fayad, R. Khatoun, S. Zeadally, and Y. Begriche, "A lightweight ECC-based authentication scheme for Internet of Things (IoT)," *IEEE Systems Journal*, vol. 14, no. 3, pp. 3440–3450, 2020.
11. S. Mishra, B. Mondal, and R. K. Jha, "Lightweight Authentication Scheme Based on ECC for IoT," *SN Computer Science*, vol. 5, no. 7, p. 949, 2024.
12. J. Lee, S. Yu, K. Park, Y. Park, and Y. Park, "Secure three-factor authentication protocol for multi-gateway IoT environments," *Sensors*, vol. 19, no. 10, p. 2358, 2019.
13. H. Timouhin, F. Amounas, and M. Azrour, "New ECC-Based IoT Authentication Protocol for Securing RFID Systems," *SN Computer Science*, vol. 4, no. 6, p. 785, 2023.
14. S. Baccouri, H. Farhat, T. Azzabi, and R. Attia, "Lightweight authentication scheme based on elliptic curve El Gamal," *Journal of Information and Telecommunication*, vol. 8, no. 2, pp. 231–261, 2024.
15. E. N. Okwa, U. A. Idiong, and U. J. Asuquo, "EC-MAP: A Lightweight Elliptic Curve-based Mutual Authentication Protocol for Secure IoT-Cloud Communication," *Journal of Engineering Research and Reports*, vol. 27, no. 4, pp. 212–225, 2025.
16. H. Timouhin, F. Amounas, and M. Badiy, "Efficient ECC-based RFID authentication for enhanced IoT security," in *Recent Advances in Internet of Things Security*, CRC Press, pp. 39–46.
17. H. Timouhin, F. Amounas, and M. Azrour, "A Survey on RFID Mutual Authentication Protocols Based ECC for Resource-Constrained in IoT Environment," in *Artificial Intelligence, Data Science and Applications*, Y. Farhaoui, A. Hussain, T. Saba, H. Taherdoost, and A. Verma, Eds., Cham: Springer Nature Switzerland, 2024, pp. 195–200. https://doi.org/10.1007/978-3-031-48573-2_28.
18. A. Armando *et al.*, "The AVISPA tool for the automated validation of internet security protocols and applications," in *Computer Aided Verification: 17th International Conference, CAV 2005, Edinburgh, Scotland, UK, July 6–10, 2005. Proceedings 17*, Springer, 2005, pp. 281–285.
19. H. Timouhin, F. Amounas, and S. Bouarafa, "Faster RFID Authentication Scheme Based on ECC for Improving the Security in IoT Environment," in *The International Conference on Artificial Intelligence and Smart Environment*, Springer, 2022, pp. 433–439.
20. R. Chaganti, A. Mourade, V. Ravi, N. Vemprala, A. Dua, and B. Bhushan, "A Particle Swarm Optimization and Deep Learning Approach for Intrusion Detection System in Internet of Medical Things," *Sustainability*, vol. 14, no. 19, p. 12828, Oct. 2022, https://doi.org/10.3390/su141912828.

Automated Intrusion Detection System Leveraging Dynamic Computational Intelligence Algorithms

Khadija Bouzaachane, El Mahdi El Guarmah, Abdullah M. Alnajim, and Sheroz Khan

Abstract The rapid proliferation of the Internet of Things (IoT) and the emergent of sophisticated cyber-attacks demonstrate the urgent need for complete, intelligent intrusion detection systems to guarantee security. Machine learning (ML) and deep learning (DL) models provide a strong groundwork for the development of such systems, as they have demonstrated an impressive performance in numerous applications, such as network intrusion detection. This chapter proposes an ML-DL approach based on the NF-UQ-NIDS-v2 dataset, which is novel and comprehensive. In this approach, ML algorithms were used for binary intrusion classification and DL algorithms were applied to perform multi-class classification, allowing for a finer-grained detection of different intrusion types. In the presence of class imbalance, the oversampling and data augmentation techniques were employed to ensure fair representation. The proposed hybrid ML-DL system demonstrates notable improvements in the accuracy and reliability of intrusion detection, representing a significant step forward for the field of cybersecurity.

Keywords Network intrusion detection systems (NIDS) · NF-UQ-NIDS-v2 dataset · K-means · Smote · Ensemble learning · Deep learning

---

K. Bouzaachane (✉)
Faculty of Sciences and Technology, Department of Computer Sciences, L2IS, Cadi Ayyad University, Marrakech, Morocco
e-mail: k.bouzaachane@uca.ac.ma

E. M. El Guarmah
Royal Air School of Aeronautics, Mathematics and Informatics Department, L2IS, Marrakech, Morocco
e-mail: guarmah@gmail.com

A. M. Alnajim
Department of Information Technology, College of Computer, Qassim University, Buraydah 51452, Saudi Arabia
e-mail: najim@qu.edu.sam

S. Khan
Department of Electrical Engineering, College of Engineering Information Technology, Onaizah Colleges, Onaizah 56447, Saudi Arabia
e-mail: cnar32.sheroz@gmail.com

M. Azrour and A. Alabdulatif (eds.), *AI-Driven Security for Next-Generation IoT Systems*, https://doi.org/10.1007/978-3-032-08784-3_4

1 Introduction

The use of networks as one of the most promising developments across various application domains, particularly IoT-Healthcare, has generated a gigantic volume of data. Connected healthcare ecosystems are designed through the integration of information technologies and biomedical engineering to ensure optimal resource utilization and address the shortcomings of traditional healthcare systems [1]. The adoption of mobile services has experienced exponential growth, enabling remote monitoring and management of a substantial number of medical devices and equipment in a compliant manner, via specially designed applications. Consequently, this has created cybersecurity risks whose severity is proportional to the volume of processed data [2]. Attackers can exploit software vulnerabilities to access networks and systems, compromising the continuous flow of network traffic by circumventing machine learning-based detectors. These types of attacks can lead to the theft of sensitive medical data for fraudulent purposes, causing considerable harm both to patient confidentiality and the reputation of healthcare institutions.

Faced with the constant evolution of cybersecurity threats, Network Intrusion Detection Systems (NIDS) have proven to be a priority research area as an essential component of developing defensive strategies. Traditional signature-based NIDS are effective at identifying known attacks but lack the flexibility to adapt to the perpetually evolving tactics of cybercriminals, who primarily target healthcare companies, hospitals, telemedicine systems, implantable medical devices, and patient databases [3]. Besides establishing guidelines, policies, and regulations to better secure connected healthcare infrastructures, specific security measures are emerging in the areas of capacity building and technical aspects of security, requiring rigorous authentication of the immense quantity of medical data.

Thus, there is growing interest in more advanced techniques, particularly those based on Artificial Intelligence [4], as a means to address the challenges posed by cyber-attacks of equivalent complexity and to protect confidential medical information against cybercrime. Computational intelligence methods constitute powerful tools equipped with intelligent architectural platforms capable of recognizing network traffic patterns and identifying irregularities indicating malicious activity. These techniques can leverage valid and comprehensive datasets, allowing them to discover new attack vectors adaptable to emerging threats. However, the effectiveness of NIDS depends on the relevance and quality of training data [5]. Existing datasets often lack representativeness of contemporary attack signatures and traffic characteristics in the medical context. This hinders the ability of models to generalize and function optimally in real scenarios of healthcare environments [5].

This chapter contributes to intrusion detection by conducting a comprehensive exploratory analysis of the NF-UQ-NIDS-v2 dataset using K-means clustering for dimensionality reduction, performing a comparative analysis of machine learning algorithms including Decision Tree (DT), Random Forest (RF), Gradient Boosting Trees (GBT), Adaboost and Extreme Gradient Boost (XG-Boost), for binary classification of network traffic anomalies, and implementing a diverse set of deep learning architectures such us Gated Recurrent Units (GRUs), Long Short-Term Memory (LSTMs), Deep Belief Networks (DBNs), Deep Neural Networks (DNNs), and Autoencoders, for refined multi-class classification of intrusions, thus offering a high-performance hybrid approach for detecting and classifying cyber-attacks in connected healthcare infrastructures.

The effectiveness of these techniques is evaluated using standard metrics including accuracy, precision, recall, F_1-score, and AUC-ROC, with particular emphasis on their ability to accurately identify malicious traffic while minimizing both false positives and false negatives. The experimental results demonstrate the superior performance of our hybrid approach compared to traditional methods, particularly in developing adaptable NIDS for healthcare environments.

The remainder of this chapter is structured as follows: Sect. 2 reviews the relevant literature on Network Intrusion Detection Systems (NIDS) developed using machine learning and deep learning techniques, focusing on research published between 2020 and 2024. Section 3 describes our methodology, including a detailed description of the evaluation metrics employed and provides an in-depth exploration of the NF-UQ-NIDS-v2 dataset. Section 4 presents a discussion of the experimental results. Finally, Sect. 5 concludes the chapter with a summary of the main findings and potential directions for future research.

2 Background and Literature Review

A plethora of recent studies have investigated computational intelligence algorithms to enhance the intrusion detection rates. The following sub-sections delve into research conducted between 2020 and 2024. The significance of these studies is discussed based on the specific type of solution proposed.

2.1 Network Intrusion Detection Systems

Intrusion detection can be defined as the monitoring of events occurring in a computer system or network, followed by an in-depth analysis of these events [6]. The objective of this analysis is to identify any indications of incidents, that is to say, violations or imminent threats of violations of IT security policies.

There are three main types of intrusion detection system such as, Anomaly based intrusion detection system, signature-based intrusion detection systems, and hybrid-based intrusion detection system [6]. Anomaly based intrusion detection system is the system that can monitor data in the real-time and detect when there are deviations from normality. This goal is accomplished using cutting-edge anomaly detection algorithms, the combination of which enables a profound data analysis with the ability to effectively recognize anomalies. This method provides an important advantage for increasing the security in environments facing a constantly evolving threat scenario.

2.2 NIDS with Machine Learning

Automatic learning has become a very powerful tool for developing effective NIDS models to detect network security intrusions. In [7] the authors aim to classify intrusions using Random Forest (RF), Linear Discriminant Analysis (LDA), Classification and Regression Trees (CART). The effectiveness of this work has been measured on the dataset KDD-CUP and compared with recent developments in the field.

The authors in [8–11] have used ten different popular machine Learning (ML) algorithms, including Decision Tree (DT), Random Forest (RF), Support Vector Machine (SVM), K-means, and Expectation–Maximization (EM) for efficient network security and computer anomaly-based intrusion detection system (AIDS). These algorithms have been tested using the most recent highly imbalanced multi-class dataset CICIDS2017 that simulates real network attacks. The K-Nearest Neighbor (KNN)-AIDS, DT-AIDS, and NB-AIDS models performed better and demonstrated greater ability to detect web attacks.

The model presented in the study [12] utilizes the UNSW-NB15 dataset that has not only achieved superior accuracy compared to previous research works, but also has effectively identified all attack categories. To enhance the performance of classifiers, the research addresses the problem of class imbalance by using SMOTE. Subsequently, the classification techniques of Decision Tree (DT), Random Forest (RF), Logistic Regression (LR), K-Nearest Neighbors (KNN), and Artificial Neural Network (ANN) have been employed.

2.3 NIDS with Deep Learning

Deep learning algorithms have shown unrivaled potential for Network Intrusion Detection Systems (NIDS). For instance, the authors in [13] have compared Recurrent Neural Networks (RNNs), Convolutional Neural Networks (CNNs), restricted Boltzmann machines (RBM), Deep Neural Networks (DNNs), Deep Belief Networks (DBNs), and Deep Autoencoders (DA) by using the CSE-CIC-IDS2018 dataset and the Bot-IoT. Their results indicated strong performance across these models, with accuracy exceeding 90%. The authors in [14] tackled network intrusion detection with imbalanced data. They used GANs to generate more data, then applied a combined CNN and BiLSTM model for detection. Testing on the CIC-IDS 2017 dataset, their proposed GAN-CNN-BiLSTM approach showed significantly better accuracy and efficiency compared to traditional methods (like Random Forest, Decision Tree) and other models (SVM, DBN, CNN, BiLSTM).

The paper [15] introduced a hybrid model using both CNNs and a Transformer encoder. After applying under-sampling and oversampling to balance the data, the CNN extracted spatial features while the Transformer captured temporal dependencies. This combined approach aimed to improve detection accuracy and minimize errors (FPs/FNs). Experiments on the NSL-KDD and CICIDS2017 benchmarks demonstrated superior performance, with higher accuracy and fewer false positives compared to existing leading models.

2.4 NF-UQ-NIDS-V2 Dataset

The NF-UQ-NIDS-v2 dataset has been constructed purposely for training and evaluating Network Intrusion Detection Systems (NIDS) [16]. The dataset is publicly available for research purposes, allowing researchers to develop and test new intrusion detection techniques. A brief overview of the research methodology used to investigate the NF-UQ-NIDS-V2 dataset has been provided. The authors in [17] suggested a hybrid model combining an Autoencoder (AE) and a Multi-Layer Perceptron (MLP) for DDoS attack detection. The AE first extracts important features from network traffic, and then the MLP uses this information to classify threats. This AE-MLP approach aims for better accuracy and fewer false alarms. During tests on the recent NF-UQ-NIDS-v2 dataset, the model's performance yielded an accuracy rate of 99.98%. The results suggest that this hybrid method performs better than several comparable techniques.

These studies have demonstrated that NIDS remain an active area of research, along with the growing adoption of computational intelligence techniques, such as Random Forest, Decision Trees, Support Vector Machines, K-Nearest Neighbor and Convolutional Neural Networks, Recurrent Neural Networks, Long Short-Term Memory networks. They have emphasized the importance of addressing data imbalance and reducing reliance on labeled data for practical implementation reasons. Furthermore, because it is newer and more balanced than previous datasets, the NF-UQ-NIDS-v2 dataset has been recognized as an invaluable tool for training and evaluating NIDS models.

3 Hybrid Approach Methodology

3.1 Introduction to the NF-UQ-NIDS-V2 Dataset

The NF-UQ-NIDS-v2 dataset is a collaborative dataset for NIDS research that includes data records from four different sources: NF-UNSW-NB15, NF-BoT-IoT, NF-TON-IoT, and CSE-CIC-IDS2018. The NF-UNSW-NB15 dataset, classifies network activity into nine classes, Benign, Fuzzers, Analysis, Backdoor, DoS,

Exploits, Generic, Reconnaissance, Shellcode, Worms. Appended to the IoT security subspace the NF-BoT-IoT the constituent provides NetFlow records derived from packet captures and classified in benign and four attack categories (Reconnaissance, DDoS, DoS, Theft). The dataset further incorporates NF-ToN-IoT, which expands the threat landscape with additional attack vectors like Injection, Password, Ransomware, Scanning and XSS. Complementing these resources, the NF-CSE-CICIDS2018 element provides NetFlow information derived from the original CSE-CIC-IDS2018 captures, featuring BruteForce attempts, Bot activities, DoS, DDoS, Infiltration, and Web Attacks.

This unified repository contains approximately 76 million network logs, with malicious traffic representing 66.88% of entries and legitimate traffic accounting for the remaining 33.12%. The distribution across specific attack categories is elaborated in Table 1. By amalgamating these diverse datasets, NF-UQ-NIDS-v2 creates a robust foundation for evaluating intrusion detection systems across varied network configurations and attack scenarios.

Table 1 Distribution of attack categories in the original dataset

Class	Count
Worms	164
Shellcode	1427
Analysis	2299
Theft	2431
Ransomware	3425
MITM	7723
Generic	16,560
Backdoor	18,978
Fuzzers	22,310
Exploits	31,551
Infiltration	116,361
Brute force	123,982
Bot	143,097
Injection	684,897
Password	1,153,323
XSS	2,455,020
Reconnaissance	2,633,778
Scanning	3,781,419
Dos	17,875,585
DDoS	21,748,351
Benign	25,165,295

Table 2 Sample of unique and common attack classes within the NF-UQ-NIDS-v2 dataset, for benign and DDos class, categorized by their source datasets

Class	Original dataset	Number of instances	Total number of instances
Benign	NF-UNSW-NB15-v2	2,295,222	25,165,295
	NF-BoT-IoT-v2	135,037	
	NF-ToN-IoT-v2	6,099,469	
	NF-CSE-CIC-IDS2018-v2	16,635,567	
DDoS	NF-BoT-IoT-v2	18,331,847	21,748,351
	NF-ToN-IoT-v2	2,026,234	
	NF-CSE-CIC-IDS2018-v2	1,390,270	

3.2 Data Preprocessing

3.2.1 Class Management

To manage computational demands of the large NF-UQ-NIDS-v2 dataset, each constituent dataset (NF-UNSW-NB15-v2, NF-BoTIoT-v2, NF-ToN-IoT-v2, NF-CSE-CIC-IDS2018-v2) was processed independently. Attack classes were categorized as unique or shared, with shared classes consolidated as in Table 2.

To address class overlap issues mentioned in [18], the "Exploit" class was excluded due to diverse attack instances. Furthermore, the "Password" class, encompassing both brute-force and network sniffing attacks [19, 20], was also removed due to heterogeneity, especially the inclusion of network sniffing attacks not represented elsewhere and differing significantly from brute-force methods. Web attacks, exploiting application layer vulnerabilities, were categorized under "Web Attack," including Injection, XSS, Infiltration, Fuzzers, Generic, and Analysis, based on OWASP's top ten critical web vulnerabilities [21].

3.2.2 Instance Sampling

The dataset was first subjected to Mini-Batch K-Means Clustering, a computationally efficient version of the K-Means algorithm, to obtain a representative sample for each class in order to expedite model training. This is accomplished by the algorithm's ability to process data in smaller, more manageable subsets. In particular, at each iteration, only a portion of the entire dataset is loaded into random-access memory (RAM), which significantly reduces the execution time of the algorithm and enhances the effective utilization of computational resources. Furthermore, instances from discrete CSV files, each comprising homogeneous class instances, were selected employing cluster sampling. The application of this probabilistic sampling method involved the grouping of original instances into clusters according to their Euclidean

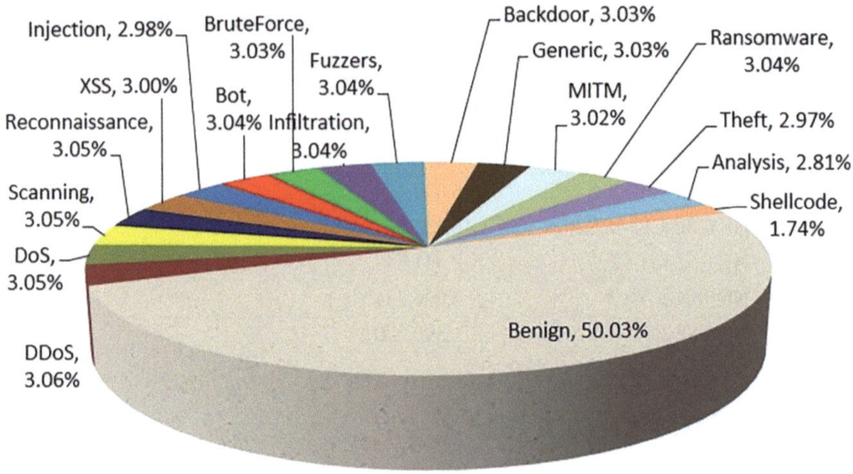

Fig. 1 K-means clustering-based sampling to create novel classes

distances, and the subsequent extraction of instances from each cluster in proportion to their relative sizes.

In our practical implementation, we organized CSV files by attack classes in separate folders before merging them into a Data Frame structure. To illustrate our approach, the Denial of Service (DoS) class, comprising over 17 million instances, was processed using the Mini-Batch K-Means algorithm from scikit-learn to identify cluster centroids and group similar data points.

We adapted the number of clusters based on dataset size: 1000 clusters for datasets exceeding 20,000 instances, 500 for those below 20,000, 250 for those under 10,000, and 125 for sets with fewer than 5000 instances. The "random-state" parameter was empirically set to 0.3. This approach allowed us to significantly reduce data volume while preserving representativeness. For example, the DoS dataset was reduced to approximately 2500 instances. This clustering-based sampling process was systematically applied to each class, resulting in a more balanced class distribution, as shown in Fig. 1.

3.2.3 Feature Normalization

Following the sampling process, we normalized all features to address the scale disparities among different attributes. This crucial step ensures that each feature contributes proportionally to the classification process regardless of its original magnitude. We employed min–max normalization using the Eq. (1):

$$x_{\text{scaled}} = \frac{x - x_{\text{min}}}{x - x_{\text{max}}} \tag{1}$$

Table 3 Distribution of attack categories after applying Mini-Batch k-means clustering

Class	Count
Benign	41,002
Web attacks	14,667
DDos	2506
Reconnaissance	2498
DOS	2497
Ransomware	2493
Bot	2489
Brute force	2485
Backdoor	2485
Scanning	2500
MITM	2471
Theft	2431
Shellcode	1427

where x represents the original feature value, and x_{min} and x_{max} denote the minimum and maximum values of that feature. This technique scales all attributes to a uniform range between 0 and 1, creating a standardized feature space that prevents attributes with naturally larger values from dominating the analysis.

3.2.4 Handling Class Imbalance (for Multi-class Classification)

In order to deal with class-imbalance of the intrusions distribution, we used the SMOTE technique to over-sample the dataset [22]. One of the difficulties encountered with sampling was the imbalance against the "Shellcode" class, as shown in Table 3. In order to prevent biasing the model from being able to distinguish "non-Shellcode" examples from "Shellcode" examples, we explicitly used SMOTE to generate synthetic "Shellcode" examples, resulting in a more balanced class-loss distribution as we show in Fig. 2. This focused oversampling strategy will help prevent the model from being biased toward the majority class samples, and therefore increases detection.

3.2.5 Proposed Hybrid System Architecture

In this chapter, we propose a hybrid Network Intrusion Detection System (NIDS) (see Fig. 3) that holds significant potential for real-world deployment in critical sectors, notably healthcare organizations reliant on interconnected systems. These environments are prime targets for cyber-attacks where disruptions can have severe consequences. Our system's architecture combines the efficiency of machine learning (ML) for initial threat detection with the nuanced classification capabilities of deep

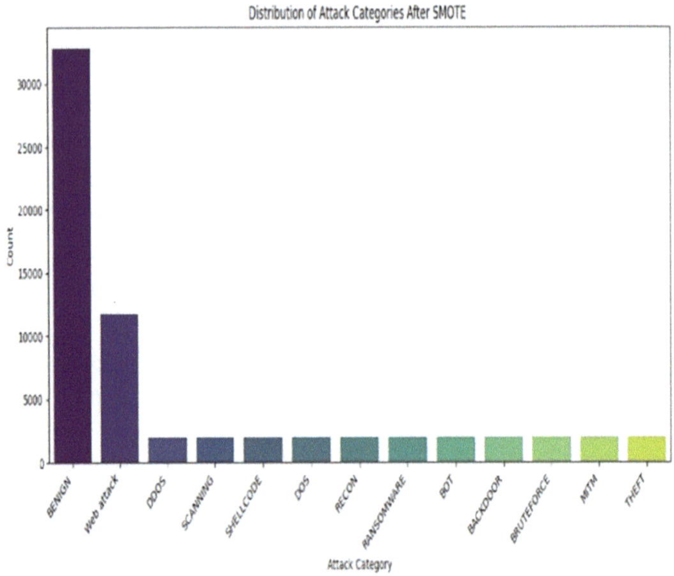

Fig. 2 Class distribution after SMOTE application

learning (DL) for identifying specific attack types. There are two crucial steps in our approach:

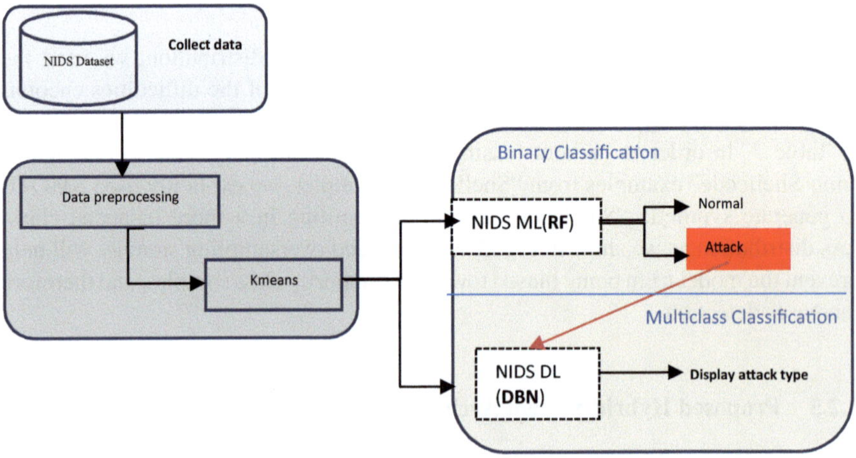

Fig. 3 Proposed NIDS security framework

- Step 1: Binary Classification, which implements Random Forest (RF), the best-performing algorithm among several prominent machine learning (ML) algorithms that we evaluated on the NF-UQ-NIDS-v2 dataset. The comparative analysis included Support Vector Machine (SVM), Decision Trees (DT), AdaBoost, XGBoost, Gradient Boosting (GB), and Logistic Regression (LR). RF exhibited superior effectiveness in binary classification for distinguishing between normal and anomalous network traffic.
- Step 2: Multi-class Classification, which employs Deep Belief Network (DBN) to further detect specific intrusion types. This algorithm outperforms the diverse set of deep learning (DL) architectures that we trained and tested for multi-class intrusion classification, including Gated Recurrent Unit (GRU), Long Short-Term Memory (LSTM), Deep Neural Network (DNN), and Autoencoder.

Indeed, this hybrid approach operates sequentially, with the Random Forest classifier first performing rapid binary classification to efficiently filter normal traffic from potential threats. Once anomalous traffic is identified, the Deep Belief Network then conducts a more computationally intensive multi-class analysis exclusively on these suspicious packets, determining the specific attack type with high precision. This two-tier architecture optimizes computational resources while maintaining detection accuracy, as the more complex deep learning analysis is applied only to the subset of traffic previously flagged as potentially malicious.

4 Experimentation and Results

4.1 Experimental Setup

To assess accurately the performance of the proposed models, it is imperative to employ instances that were not utilized during the training process. This requires the division of our dataset of 81, 951 into two distinct subsets: a training set of 80% of the overall dataset and a test set of 20% of the overall dataset. The size of the training data is 65,560 and the size of the testing data is 16,391. Training of our proposed models has been implemented on a workstation with the following specifications: Intel(R) Core(TM) i7-6600U, CPU@2.60 GHz (4 CPUs), 16Go in RAM memory, and we used for this purpose Python programming language and the Scikit-learn, keras, Numpy, and pandas libraries.

4.2 Evaluation Metrics

Quantitative measures are essential for a thorough evaluation of algorithms [23]. In this study, we used a combination of measures such as precision, accuracy, recall, F_1-score, macro-average AUC-ROC, micro-average AUC-ROC, learning and prediction time and computational cost to evaluate the performance of the aforementioned algorithms applied to the NF-UQ-NIDS-v2 database.

4.3 Binary Classification Results

This section evaluates several machine learning algorithms for binary attack classification. The models tested include LR, SVM, DT, RF, AdaBoost, GBT, and XGBoost. The dataset used for this task features a balanced distribution between classes. Before training, the dataset was thoroughly shuffled to help ensure unbiased model learning and improve the models' ability to generalize to new data. This shuffling was performed using the shuffle function available in the Keras library.

We trained each model using fixed hyperparameters. For the SVM, the regularization parameter was set to 50 to balance the classification margin and errors. We used the default settings for the LR and DT models. For RF, the number of trees was set to 11. For the boosting algorithms, the number of boosting stages was set to 100 forGBT, 100 for AdaBoost, and 240 for XGBoost. As detailed in Table 4, the training process resulted in the following accuracies: 0.99 for SVM, 0.86 for LR, 1.0 for DT, AdaBoost, and RF, and 0.99 for GBT and XGBoost.

The performance metrics summarized in Table 5 clearly show that LR performed significantly worse than the other models, achieving only 85% across accuracy, precision, recall, and F_1-score. More critically, LR resulted in a very high number of false negatives (1129 instances), meaning it frequently misclassified actual attacks as benign traffic. This level of performance is inadequate for a reliable intrusion detection system.

In contrast, all other evaluated models demonstrated excellent performance. SVM, DT, AdaBoost, GBT, and XGB all achieved accuracy, precision, recall, and F_1-scores

Table 4 Summary of training process

Models	Training time (s)	Accuracy
SVM	1.44427	0.99
LR	0.00417	0.86
DT	1.49690	1.0
RF	0.01253	1.0
GB	0.29613	0.99
Adaboost	0.28366	1.0
XGboost	0.01257	0.99

Table 5 Summary of results of testing process

	LR	SVM	DT	RF	Adaboost	GBT	XGB
Time of prediction (s)	0.010	6.025	0.0109	0.031	0.848	0.08	0.045
Accuracy	0.85	0.998	0.999	1.0	0.999	0.999	0.9999
Precision	0.858	0.998	0.999	1.0	0.999	0.999	0.999
Recall	0.858	0.998	0.999	1.0	0.999	0.999	0.999
F_1-score	0.858	0.998	0.999	1.0	0.999	0.999	0.999
FN count	1129	12	3	0	1	0	0

of 99.8 or 99.9%. While highly effective, these models still produced a small number of false negatives (ranging from 0 for RF to 3 for DT). Random Forest (RF) stood out as the top performer, achieving perfect scores across all metrics: 100% accuracy, precision, recall, and F_1-score. Crucially, RF resulted in zero false negatives, meaning it correctly identified every single attack instance in the test set, Fig. 4.

When considering computational efficiency (Tables 4 and 5), LR offered the fastest training time (0.00417 s) but its poor accuracy makes this irrelevant. SVM was notably slow to train (1.44427 s). RF exhibited moderate training (0.01253 s) and prediction times (0.031 s), which are practical for real-world applications where high accuracy and reliability are paramount.

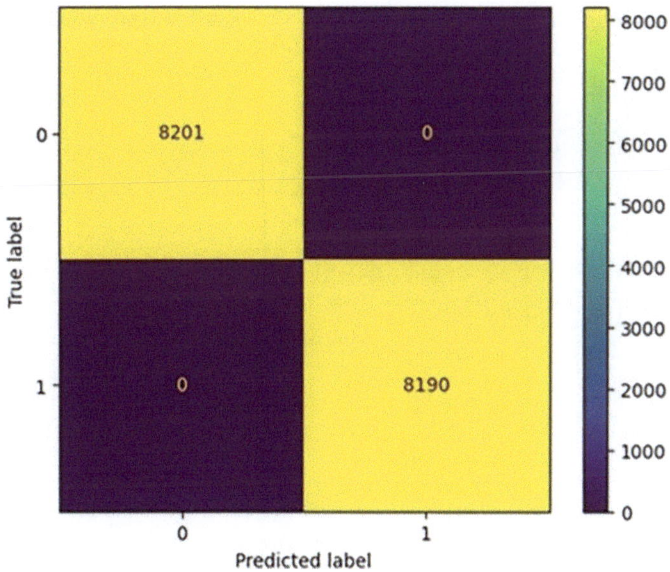

Fig. 4 Confusion matrix of RF

4.4 Multi-class Classification

This section evaluated five different deep learning models—LSTM, GRU, DNN, DBN, and Autoencoder—for the complex task of classifying network traffic into 13 distinct categories, including various attack types and benign traffic.

During training, we monitored both training and validation accuracy to guide the learning process for each model (Fig. 5). In the final testing phase, the Deep Belief Network (DBN) model clearly stood out. It achieved the highest test accuracy at 99% (Fig. 6a), significantly outperforming the other four models evaluated.

Furthermore, the Receiver Operating Characteristic (ROC) curve analysis provided further strong evidence of the DBN's effectiveness (Fig. 6b). Both the micro-averaged and macro-averaged Area Under the Curve (AUC) scores were nearly

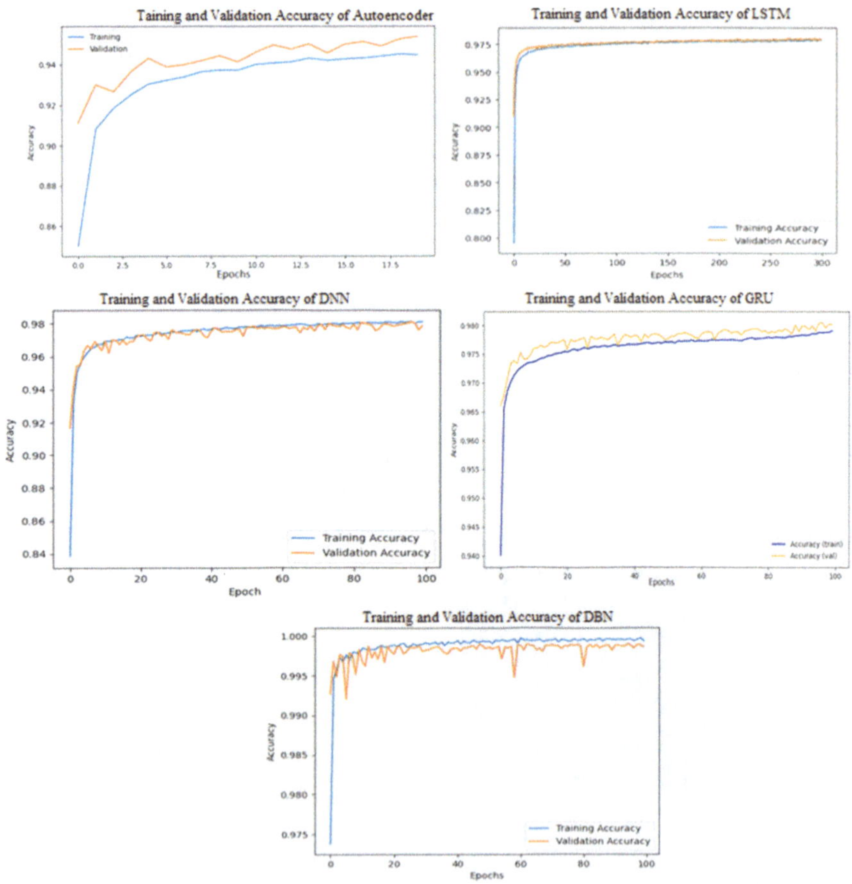

Fig. 5 Training and validation accuracy of DL models

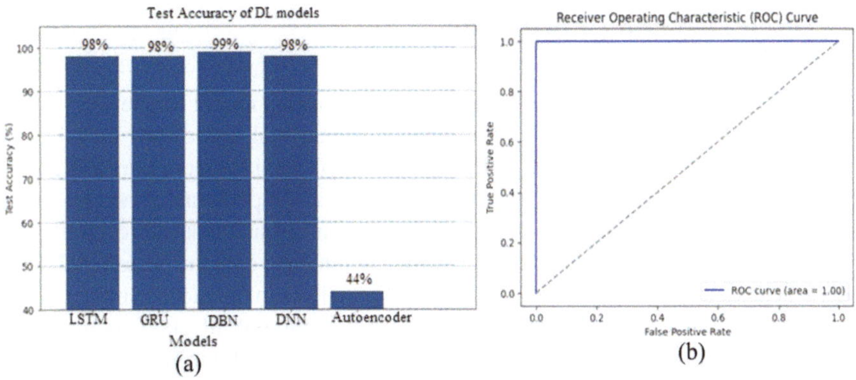

Fig. 6 **a** DL model performance: test accuracy comparison; **b** ROC curve analysis of DBN model for attack type

perfect at 0.9996. This highlights the DBN's outstanding ability to reliably distinguish between different attack types and normal traffic with very high sensitivity and specificity.

We assessed the computational characteristics of each model (summarized in Table 6). This revealed significant differences:

- The Autoencoder model was the fastest to train (134 s) and offered instant predictions (1 s), but had the highest number of parameters (32,503), indicating greater model complexity.
- The Deep Belief Network (DBN), after its pre-training phase, resulted in a final classifier with the fewest parameters (10,765). It also achieved very fast predictions (1 s), with a moderate training time (527 s).
- Recurrent models, LSTM and GRU, showed considerable complexity (28,237 and 21,389 parameters) and required the longest training times (587 s and 540 s, respectively), though their prediction speeds remained quick (3–4 s).
- The standard Deep Neural Network (DNN) had moderate complexity (14,605 parameters) and fast predictions (2 s), but unexpectedly took the longest to train in our setup (660 s).

Table 6 Performance comparison of deep learning models for multi-class attack classification

DL models	GRU	LSTM	DBN	DNN	Autoencoder
Training time (s)	540	587	527	660	134
Time of prediction (s)	4	1	3	2	1
Computational cost (parameters)	21,389	28,237	10,765	14,605	32,503

5 Discussion and Conclusions

This study aimed to identify an effective machine learning model for Network Intrusion Detection Systems (NIDS). After evaluating several algorithms on NF-UQ-NIDS-v2 dataset, we selected a hybrid approach combining Random Forest (RF) and Deep Belief Network (DBN). The results showed that RF performs very well on the initial task of binary classification (separating normal from attack traffic), achieving high accuracy "100%." This efficient filtering step means only suspected malicious traffic is passed to the DBN model. The DBN specializes in the more complex multi-class classification needed to determine the specific type of attack. This two-stage architecture leverages the RF's speed and accuracy for initial filtering, and the DBN's capability to model complex relationships for detailed attack categorization.

Despite its potential, the hybrid RF-DBN method involves inherent limitations that must be carefully addressed for effective real-world deployment. Error propagation stands out as the most critical limitation. Specifically, if the initial Random Forest (RF) incorrectly flags benign traffic as an attack (a False Positive), it triggers unnecessary processing by the Deep Belief Network (DBN). This can lead to false alerts if the DBN fails to correct the initial misclassification. To mitigate this, one practical approach is to train the DBN specifically on the False Positives generated by the RF, enabling it to potentially rectify errors from the first stage.

Furthermore, the issue of interpretability arises. While Random Forests are already less transparent than simple decision trees, DBNs are notoriously opaque "black boxes." Consequently, understanding precisely why the system classified a particular network flow as a specific type of attack becomes difficult. This lack of clarity complicates the analysis of security alerts and can hinder trust in the system's decisions. To address this challenge, tools like SHAP or LIME can be employed. These techniques provide local explanations for the model's outputs by identifying the features that most influenced a given prediction.

Another significant concern relates to the system's ability to generalize to new, previously unseen attacks (Zero-Day threats). Because the model is trained only on known attack types present in the dataset, it may struggle to correctly identify entirely novel threats or significant variations of existing attacks that were not part of its training experience. Several strategies can help address this:

- Firstly, implementing a process for regularly retraining both models using recent network data that includes newly identified threats is crucial.
- Secondly, supplementing the RF-DBN architecture with an unsupervised model, such as an Autoencoder, can be beneficial. Such models excel at detecting highly unusual behavior, even if it doesn't match any known attack signature, effectively acting as a safety net.
- Finally, continuous monitoring of key performance indicators, like detection rates and false alert frequencies, is essential. Any degradation in performance should be investigated promptly, as it might signal the emergence of new, undetected threats.

Although, we used SMOTE to address the dataset's initial class imbalance, we need to consider how the model would perform with *severely* unbalanced data. In real-world situations, some critical attacks are extremely rare, creating much greater imbalance than addressed in this study. This extreme rarity poses a challenge: the initial Random Forest stage might fail to detect these infrequent attacks, and even if they reach the Deep Belief Network, learning their patterns effectively can be difficult, despite oversampling. Consequently, the system's ability to detect these rare but important threats could be reduced. To handle such scenarios better, future work could explore more advanced strategies beyond standard SMOTE. Options include adaptive sampling techniques (like ADASYN), cost-sensitive learning (which gives more weight to errors on rare classes), or ensemble methods specifically designed for imbalanced data. Evaluating the model under these demanding, highly imbalanced conditions is crucial for ensuring its reliability in real-world network environments.

In conclusion, while the hybrid RF-DBN model demonstrates significant potential for intrusion detection due to its specialized two-stage structure, its successful deployment hinges on acknowledging and proactively managing its inherent limitations. Implementing practical solutions, such as those suggested for error propagation, interpretability, and zero-day detection, will be vital for building a robust, reliable NIDS capable of adapting to the constantly evolving threat landscape.

References

1. Mrinai M, Shailaja D, Patil C, Internet of Things: A survey of enabling technologies in healthcare and its applications, Computer Networks, Volume 153, Pages 113–131, https://doi.org/10.1016/j.comnet.2019.03.006 (2019)
2. Mahesh M, Gadekar A, and Gadhade Y: Internet of Things (IoT): a survey. In 2019 IEEE pune section international conference (PuneCon), pp. 1–6. IEEE, https://doi.org/10.1109/PuneCon46936.2019.9105831(2019)
3. Qiu H, Qiu M, Liu M and Memmi G: Secure Health Data Sharing for Medical Cyber-Physical Systems for the Healthcare 4.0, in IEEE Journal of Biomedical and Health Informatics, vol. 24, no. 9, pp. 2499–2505, Sept. https://doi.org/10.1109/JBHI.2020.2973467 (2020)
4. Vourganas I. J and Michala A.L: Applications of Machine Learning in Cyber Security: A Review.J. Cybersecur. Priv. 4(4), 972–992. https://doi.org/10.3390/jcp4040045 (2024)
5. Al-Riyami S, Lisitsa A, Coenen F: Cross-Datasets Evaluation of Machine Learning Models for Intrusion Detection Systems. Proceedings of Sixth International Congress on Information and Communication Technology. Lecture Notes in Networks and Systems, vol 217. Springer, Singapore. https://doi.org/10.1007/978-981-16-2102-4_73 (2022)
6. Neha K: Cybersecurity in healthcare: a narrative review of trends, threats and ways forward (2024). Journal of Cyber Security. https://doi.org/10.6084/m9.figshare.25346236.v1 (2024)
7. Scarfone K. A, and Mell P. M: Guide to Intrusion Detection and Prevention Systems (IDPS), National Institute of Standards and Technology, Gaithersburg, MD, NIST SP 800-94. https://doi.org/10.6028/NIST.SP.800-94 (2007)
8. Saranya T, Sridevi S, Deisy C, Chung TD, Khan MA: Performance analysis of machine learning algorithms in intrusion detection system: A review. Procedia Comput. Sci.171: 1251–1260. https://doi.org/10.1016/j.procs.2020.04.133 (2020)
9. Maseer ZK, Yusof R, Bahaman N, Mostafa SA, Foozy CFM. Benchmarking of machine learning for anomaly-based intrusion detection systems in the CICIDS2017 dataset. IEEE Access, 9, 22351–22370. https://doi.org/10.1109/ACCESS.2021.3056614 (2021)

10. Al Lail M, Garcia A and Olivo S: Machine Learning for Network Intrusion Detection—A Comparative Study. Future Internet, 15(7), 243. https://doi.org/10.3390/fi15070243 (2023)
11. Stiawan D, Idris M.Y.B, Bamhdi A.M, Budiarto R: CICIDS-2017 dataset feature analysis with information gain for anomaly detection. IEEE Access, 8, 132911–132921. https://doi.org/10.1109/ACCESS.2020.3009843 (2020)
12. Elmrabit, N, Zhou F, Li F, Zhou H: Evaluation of machine learning algorithms for anomaly detection. In Proceedings of the 2020 International Conference on Cyber Security and Protection of Digital Services (Cyber Security), Dublin, Ireland, 15–19 June; pp. 1–8. https://doi.org/10.1109/CyberSecurity49315.2020.9138871(2020)
13. Ahmed HA, Hameed A, Bawany NZ: Network intrusion detection using oversampling technique and machine learning algorithms. PeerJ Computer Science 8: e820. https://doi.org/10.7717/peerj-cs.820. (2022)
14. Ferrag MA, Maglaras L, Moschoyiannis S, and Janicke H: Deep Learning Approaches for Network Intrusion Detection: A Performance Evaluation. Journal of Information Security and Applications. 50. https://doi.org/10.1016/j.jisa.2019.102419 (2020)
15. Li S, Li Q, Li M: A Method for Network Intrusion Detection Based on GAN-CNN-BiLSTM. International Journal of Advanced Computer Science and Applications, Vol. 14, No. 5, https://doi.org/10.4018/IJDCF.2019070104 (2023)
16. Chen H, Gui-Rong Y, Yeou-Ren S: Hybrid Intrusion Detection System Based on Data Resampling and Deep Learning. (IJACSA) International Journal of Advanced Computer Science and Applications, Vol. 15, No. 2. https://doi.org/10.14569/IJACSA.2024.0150214 (2024)
17. Sarhan M, Layeghy S, and Portmann M: Towards a Standard Feature Set for Network Intrusion Detection System Datasets, Mobile Networks and Applications, 103, 108379. https://staff.itee.uq.edu.au/marius/NIDSdatasets/. [Accessed 01–10–2023] (2022)
18. Adeniyi O, Sadiq A.S, Pillai, P, Aljaidi M, Kaiwartya O: Securing Mobile Edge Computing Using Hybrid Deep Learning Method. Computers, 13, 25. https://doi.org/10.3390/computers13010025 (2024)
19. Zoghi Z, and Serpen G: UNSW-NB15 computer security dataset: Analysis through visualization. Security and Privacy 7, no. 1: e331. https://doi.org/10.1002/spy2.331 (2024)
20. Derbyshire R, Green B, Prince D, Mauthe A, and Hutchison D: An analysis of cyber security attack taxonomies. In 2018 IEEE European Symposium on Security and Privacy Workshops (EuroS&PW), pp. 153–161. IEEE. https://doi.org/10.1109/EuroSPW.2018.00028 (,2018)
21. Fredj O.B, Cheikhrouhou O, Krichen M, Hamam H, Derhab A: An OWASP Top Ten Driven Survey on Web Application Protection Methods. In: Garcia-Alfaro, J., Leneutre, J., Cuppens, N., Yaich, R. (eds) Risks and Security of Internet and Systems. CRiSIS 2020. Lecture Notes in Computer Science(), vol 12528. Springer, Cham. https://doi.org/10.1007/978-3-030-68887-5_14 (2021)
22. Jiao X, Li J: An Effective Intrusion Detection Model for Class-imbalanced Learning Based on SMOTE and Attention Mechanism. Computer Science Conference on Privacy, Security and Trust. https://doi.org/10.1109/PST52912.2021.9647756 (2021)
23. Gulshan K: Evaluation Metrics for Intrusion Detection Systems-A Study. International Journal of Computer Science and Mobile Applications. 11. 2015. https://www.ijcsma.com/abstract/evaluation-metrics-for-intrusion-detection-systems-a-study-95721.html. (2015)

A Deep Q-Network-Based Detection for Mobile Edge Computing Security

Hafida Assmi, Azidine Guezzaz, Said Jabbour, Said Benkirane, and Mourade Azrour

Abstract Mobile Edge Computing (MEC) improves the performance of real-time services by positioning computational resources at the network edge, closer to users, thereby minimizing latency. However, the decentralized nature of edge computing makes its infrastructure more susceptible to cyber threats, underscoring the need for intrusion detection systems (IDS) to enhance security. This study analyzes task offload decision-making in a MEC environment, where tasks can be executed either on the edge cloud or on a mobile device. To this end, a hybrid model combining machine learning techniques (Support Vector Machine, Random Forest, Gradient Boosting) and a Deep Q-Network (DQN) neural network is proposed. Experimental results based on data from Shanghai and Guangzhou demonstrate notable improvements, with accuracy rates ranging between 67 and 99%.

Keywords Security · Decision-making · Mobile edge computing · Task offloading

1 Introduction

As heterogeneous IoT applications proliferate, the performance demands for latency and jitter are becoming more rigorous, highlighting the limitations of the traditional cloud computing model. To overcome these challenges, multi-access edge computing (MEC) has been implemented as a viable solution, placing computing

H. Assmi · A. Guezzaz (✉) · S. Benkirane
SISAR Team, LaRTID Lab, Higher School of Technology Essaouira, Cadi Ayyad University, Marrakech, Morocco
e-mail: A.GUZZAZ@gmail.com

H. Assmi · S. Jabbour
MIA laboratory, MSIA Team, Faculty of Sciences and Techniques, Moulay Ismail University of Meknes, Errachidia, Morocco

M. Azrour
IDMS Team, FacultyofSciencesandTechniques, Moulay Ismail University of Meknes, Errachidia, Morocco

M. Azrour and A. Alabdulatif (eds.), *AI-Driven Security for Next-Generation IoT Systems*, https://doi.org/10.1007/978-3-032-08784-3_5

resources closer to users. However, implementing IoT applications in a MEC environment presents significant challenges, particularly around security and privacy. Safeguarding data and maintaining confidentiality are key factors in the successful adoption of this decentralized architecture. This paper focuses on analyzing task offloading decision-making within a Mobile Edge Computing (MEC) environment, where tasks be processed either on the edge cloud or directly on the mobile device. To optimize this decision, a hybrid model is proposed, combining classical machine learning techniques, including RF, GB, and SVM with a Deep Q-Network (DQN) neural network. The latter, derived from deep reinforcement learning, combines the Q-Learning algorithm with neural networks to improve decision-making. The approach developed aims to predict the best load-shedding strategy based on key network parameters such as latency, downlink traffic and duration, using Shanghai and Guangzhou datasets. Two major contributions were highlighted:

Firstly, the DQN reinforcement algorithm is employed to enhance task offloading decisions in an edge computing environment (ECM).

Secondly, to improve decision-making and model performance, three machine learning algorithms are implemented to predict the most effective load-shedding decisions.

The remainder of this paper is organized as follows: Sect. 2 reviews the background and discusses related research on approaches to Mobile Edge Computing. Section 3 details the design of the framework proposed in this study. Section 4 discusses the experimental evaluation and highlights the results derived from the model. The paper concludes with a summary of the primary contributions and proposes avenues for future research.

2 Background and Related Works

This section reviews essential concepts and recent progress in intrusion detection systems (IDS) within the context of Mobile Edge Computing (MEC), with a focus on enhancing the security of Internet of Things (IoT) devices. MEC provides storage, computing, and application services through edge nodes, including wearable devices, autonomous vehicles, smartphones, and sensors, acting as a key driver in the evolution of the Internet of Things (IoT) [1]. The MEC architecture facilitates seamless communication between applications and services, connecting cloud computing with end users. It consists of interconnected devices and systems, structured in layers and deployed flexibly. However, as technology advances, MEC is confronted with more sophisticated and complex threats [2].

In 2025, Attou et al. [3] developed an IDS model that integrates the Radial Basis Function Neural Network (RBFNN) with gradient boosting. Their study evaluates two approaches: the first applies RBFNN directly to the NSL-KDD and CICIDS2017 datasets, while the second uses Adaboost for feature selection before training the RBFNN. The experimental results confirm the superior performance of the second approach, achieving accuracies of 99.7% and 99.5%, while significantly lowering the false alarm rate to 0.6% and 0.18%, respectively. These results underscore the superior performance of the proposed model over the standalone use of RBFNN and other existing methods.

Additionally, with Assmi et al. [4] an intrusion detection model (IDS) based on machine learning was proposed by integrating Random Forest (RF), Support Vector Machine (SVM), and K-Nearest Neighbors (KNNs) through an ensemble learning strategy using a Voting Classifier. It was evaluated on the Edge-IIoT and NSL-KDD datasets, where it delivered noteworthy performance. Experimental findings indicate that accuracy spans from 72 to 99%, with precision values ranging from 87 to 99%. Recall and F1-scores also fall within the 72% to 99% range, reflecting the model's robust capability in recognizing both normal behavior and attacks. Dargaoui et al. [5], examined the attack vectors targeting the different layers of the Internet of Things (IoT) architecture and provided an overview of various machine learning algorithm classes. They then reviewed the application of these techniques in enhancing IoT security. Finally, their work highlights the key challenges and issues related to the integration of machine learning into IoT security mechanisms. Furthermore, Belali et al. [6] investigated the escalating security issues faced by the Internet of Things due to an increase in cyberattacks, and highlighted the advantages of deep learning-based intrusion detection systems, particularly those combining convolutional neural networks with long short-term memory networks, to bolster the security of IoT networks. Dargaoui et al. [7] developed a three-factor anonymous authentication protocol based on elliptic curve cryptography, with its security validated through both informal and formal analyses. Their solution demonstrates superior security, reduced computational demands, lower computational overhead, and improved communication efficiency. Mohy-Eddine et al. [8] developed a deep learning-based intrusion detection system (IDS) that enhances performance through chi-square feature selection and autoencoder-based outlier removal. Tested on the NF-Bot-IoT and NF-ToN-IoT datasets, the model achieved remarkable accuracy of 99.84% and a strong Matthews Correlation Coefficient, showcasing its effectiveness in handling data imbalance.

In 2024, Wu et al. [9] proposed a privacy-preserving offloading strategy that applies stochastic game theory, incorporating multiple access points. This approach includes the development of a multi-access point offloading framework and a Quality of Service (QoS) model. In addition, they introduced an optimal deep reinforcement learning (DRL) algorithm, combined with privacy preservation (JODRL-PP), to optimize offloading decisions. The results validate the effectiveness of both their proposed model and algorithm. Amaouche et al. [10] aimed to protect AVENT networks from attacks like Denial of Service (DoS). To tackle this challenge, they introduced an intrusion detection system (IDS) that employs ensemble learning, leveraging the AdaBoost algorithm. The model was tested on three datasets NSL-KDD, TON-IoT, and UNSW-NB15 and showed superior performance compared to current IDS solutions. It attained near-perfect scores across all key metrics, including accuracy, precision, recall, and F1-score.

Liu et al. [11] explored the issue of integrated resource allocation and task offloading in MEC, focusing on the diverse resources offered by heterogeneous

MEC servers to manage remote tasks on mobile devices. To solve the initial aspect of the problem selecting the winning nodes they designed an incentive mechanism to ensure overall system stability. Approximation ratios were also calculated for configurations with one or multiple MEC servers. To address the second challenge, making load-shedding decisions—they designed an approximation algorithm proved its ability to deliver a polynomial-time solution in the case of a single MEC server. He results indicate that the proposed system can effectively identify optimal solutions within dynamic mobile environments. Concomitantly, Hazman et al. [12] conducted a study highlighting the critical importance of intrusion detection systems (IDS) in smart cities, focusing on their role in anomaly detection, data breach prevention, and ensuring the security and resilience of urban digital infrastructure. The study points to the necessity of regularly updating and deploying advanced IDS technologies that leverage contemporary machine learning and deep learning methods to strengthen smart city cybersecurity and safeguard citizen privacy. In addition, Bella et al. [13] developed DNDF-IDS, an intrusion detection system (IDS) utilizing Deep Neural Decision Forests. To boost its performance, they incorporated four feature selection methods: LASSO Regression (LR), Principal Component Analysis (PCA), Random Forest Feature Importance (RFFI), and SelectKBest. The model was evaluated on three datasets, NSL-KDD, UNSW-NB15, and CICIDS2017, reaching accuracy levels between 94.09% and 98.84%. Furthermore, it shows an impressive prediction time of 0.1 ms per record, ensuring rapid and effective intrusion detection. Mohy-eddine et al. [14] developed an intrusion detection system (IDS) built on Radial Basis Function Neural Networks (RBFNNs). To improve its effectiveness, they integrated several ensembles learning techniques, including LightGBM (LGBM), XGBoost (XGB), Extra Trees (ET), Random Forest (RF), AdaBoost (ADA), to identify the most important features from the datasets. The model, evaluated on the NF-Bot-IoT and NF-ToN-IoT datasets, attained an accuracy (ACC) of 99.25% and 96.92%, respectively, with Matthews Correlation Coefficients (MCC) of 82.97% and 90.05%. Dargaoui et al. [15] conducted a brief analysis of the security issues related to the Internet of Things in smart agriculture. Emphasizing the vital importance of authentication for IoT security, they subsequently surveyed the primary types of authentication mechanisms utilized in this context. In addition, Anwar et al. [16] presented a contemporary perspective on the application of blockchain and IoT in the context of Industry 4.0 and Society 5.0. Their work addresses key issues, including unresolved challenges, current limitations, and emerging research opportunities that can guide future investigations and foster broader adoption of these innovations.

In 2023, Wang et al. [17] carried out a thorough examination of security and privacy in MEC, focusing on the role of artificial intelligence (AI). Furthermore, they identified emerging security and privacy challenges and explored potential solutions through AI and other technologies. Additionally, Alzubi et al. [18] analyzed the security features of blockchain, including its transparency and resilience against attacks, among other aspects. Using a combination of theoretical analysis and experimental simulations, they showed that the proposed framework improves security, reduces latency, controls access, boosts throughput, and cuts operating costs. Ning et al. [19] investigate the combination of mobile computing and ML within the Internet of

Drones, focusing on essential technologies, solutions, and the challenges associated with Unmanned Aerial Vehicles (UAVs). The research also reviews current MEC and ML solutions within the Internet of Drones, considering the challenges identified. Lastly, they discuss several ongoing challenges and open questions in this area.

In 2022, Zhang et al. [20] investigated a MEC network integrated with an Intelligent Reflective Surface (IRS) within the context of physical layer security. In this setup, users can temporarily offload sensitive and resource-demanding processes to a computer access point (CAP), aided by the IRS. Concomitantly, Singh et al. [21] presented an extensive study on MEC technology, detailing crucial aspects such as objectives, network architecture, and more. The research also explores the incorporation of artificial intelligence (AI) and federated learning (FL) within MEC, facilitating collaboration among mobile devices to develop a shared predictive model. Additionally, it emphasizes security challenges, attack types, and privacy concerns.

In 2021, Ali et al. [2] explored the key threats to EMC systems, outlining their architecture, identifying vulnerable functional layers, and categorizing potential security risks and measures. They recommend that CEM providers employ a multi-level security framework to effectively defend against targeted attacks.

In 2020, Li et al. [22] developed an innovative authentication framework tailored for the MEC environment to ensure secure and efficient communication. Building on this, they designed an identity-based anonymous system specifically optimized for lightweight devices. The result show that this system delivers better performance than various existing solution. In addition, Rahman et al. [23] explored the concepts, features, privacy protections, and applications of privacy in MEC driven by IoT, highlighting the significance of information in the modern world. Using autonomous vehicles as a case study, they evaluated the security mechanisms of MEC within IoT systems. Furthermore, Cao et al. [24] explored edge computing, offering a comprehensive overview of its concepts while comparing it to cloud computing. Their work also covers edge computing architecture, essential technologies, security and privacy challenges, as well as the diverse applications of edge computing.

In addition, Rahman et al. [23] explored the concepts, features, privacy protections, and applications of privacy in MEC driven by IoT, highlighting the significance of information in the modern world. Using autonomous vehicles as a case study, they evaluated the security mechanisms of MEC within IoT systems. Furthermore, Cao et al. [24] explored edge computing, offering a comprehensive overview of its concepts while comparing it to cloud computing. Their work also covers edge computing architecture, essential technologies, security and privacy challenges, as well as the diverse applications of edge computing.

In 2019, Huang et al. [25] proposed SEECO, a compute offloading strategy that integrates security and energy efficiency, tailored for service workflows within the MEC environment. Extensive experimentation with various workflow parameters demonstrated SEECO's ability to improve both security and energy efficiency for mobile applications. Concomitantly, Xiao et al. [26] performed an in-depth analysis of the most significant and fundamental attacks, including DDoS attacks, malware injections, breaches, and issues related to authentication and authorization. Their study investigated the underlying causes of these issues, assessed the current security landscape in edge computing, identified critical challenges, and proposed potential avenues for future research. Zhang et al. [27] developed a task migration method for MEC systems using Deep Q-Network (DQN), enabling the model to autonomously learn the optimal migration policy from past experiences, without requiring prior knowledge of the user mobility model. This approach utilizes the temporal difference (TD) learning algorithm within the framework of a Markov decision process (MDP), to enable ongoing learning. Extensive simulations within the MEC environment validate the effectiveness of the proposed task migration approach.

In 2018, Roman et al. [28] performed a study that offers a comprehensive analysis of the threats, challenges, and security mechanisms across various edge computing paradigms, while also highlighting potential synergies and opportunities for collaboration. Their findings indicate that, in order to enhance security and efficiency, each edge computing paradigm should draw on the advancements of the others.

Simultaneously, He et al. [29] examined the security challenges related to two IoT applications that stand to benefit from MEC technology. The first is an environmental monitoring system based on an industrial IoT network, while the second, a rapidly growing area, centers around mobile IoT supported by a network of Unmanned Aerial Vehicles. Meanwhile, Zhao et al. [30] applied genetic algorithms to optimize system performance. Their approach is built upon the LODCO algorithm, an optimized dynamic offloading technique that employs the Lyapunov method. LODCO converts the time-dependent problem into a deterministic time-slot problem through Lyapunov optimization, incorporating a disturbance parameter specific to each mobile device. They proposed a multi-user, multi-server model that combines the genetic algorithm with Greedy-type strategies to achieve optimal system performance. Wu et al. [31] proposed SVMO, an optimized offloading algorithm that integrates Support Vector Machines (SVMs) with a decision tree to create training data. This strategy enables accurate and efficient subtask allocation in vehicular networks by deciding whether each subtask should be processed locally or offloaded. Simulation results show that SVMO delivers improved decision accuracy, faster convergence, and shorter response times than other competing algorithms.

3 Outlined Model

This section offers an in-depth description of the model's architecture and describes the algorithms integrated into design and development process.

3.1 Designed Architecture

This section details the process of developing a decision analysis model for task offloading in an MEC environment, with the model's architecture shown in Fig. 1.

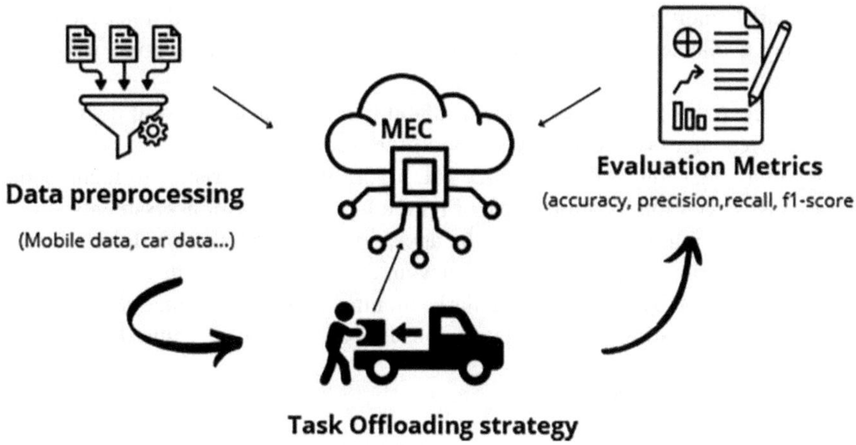

Fig. 1 Structure of the MEC-RGSD approach

Data preprocessing: The project began with comprehensive data preprocessing to guarantee data quality and confidentiality. This critical step prepared the data for analysis and modeling, resulting in a clean and secure dataset.

Task offloading strategy: This project employs machine learning and intelligent algorithms to enhance the distribution of tasks between mobile devices and edge cloud. By evaluating factors like network conditions, traffic load, and user habits to improve the performance of the MEC network.

Evaluation metrics: These metrics evaluate the efficiency of the task offloading strategy, focusing on minimizing latency and enhancing decision accuracy. They are crucial for evaluating the effectiveness of dynamic task allocation between edge cloud and mobile devices.

3.2 Details of the Proposed Model Dynamic Intelligent Edge Task Offloading

The DITO algorithm transforms task management in edge computing by intelligently and dynamically optimizing task offloading between local devices and edge or cloud processing resources, leading to enhance application efficiency and responsiveness.

Algorithm

- Task Assessment: Evaluate the tasks to be executed by considering their resource demands, complexity, and processing time.

- Network Conditions Analysis: Analyze the existing network conditions, including latency, available bandwidth, and congestion, to determine the optimal offloading options.
- Resource Assessment: Assess the availability and capacity of local devices and remote resources (cloud or edge servers) for task execution.
- Dynamic Decision-Making: Utilize ML algorithms or heuristics to identify the best task execution location (local or remote), based on prior assessments.
- Task Offloading: Offload tasks to the appropriate server (edge or cloud) when it improves performance, ensuring latency and resource constraints are met.
- Monitoring and Adaptation: Regularly monitor task performance and network conditions, adjusting offloading decisions in real-time when required.

Deep Q-Network (DQN)

DQN is a strategy employed in reinforcement learning that leverages deep neural networks to approximate the action-value function (Q function) in environments where traditional methods, such as Q-learning, may not be effective. It allows an agent to make decisions using complex inputs, such as images or unstructured data, and steadily improves through experience to optimize rewards over time.

Algorithm

- Estimating the Q function using a neural network:

The neural network $Q(s,a,\theta)$ processes the input state s and generates Q as output values corresponding to each possible action a.

The goal is to train the network to identify a policy that achieves the maximum long-term cumulative reward.

- Utilizing Experience Replay:

The observed transitions *(s, a, r, s')* are stored in a replay buffer.

During training, transitions are randomly sampled in mini-batches to minimize temporal correlations and improve training stability.

- Implementation of a Target Network (*Q'*):

A target network $Q'(s,a;\theta-)$ is employed to calculate a target for learning.

This network is a copy of the main network, with updates happening less frequently to prevent unstable oscillations.

- Modifying the Q function using the Bellman equation:

Learning advances by adjusting the estimation Q value to minimize the average difference between the estimated output and the target value, as given by:

$$y_i = \begin{cases} r_j \text{ si } s_{j+1}\text{est un état final} \\ r_j + \gamma \max_a Q'\left(s_{j+1}, a'; \theta^-\right)\text{sinon} \end{cases} \tag{1}$$

The weights θ of the network are modified by stochastic gradient descent (SGD).

- ϵ-greedy exploration method:

To promote exploration of the environment, the DQN adopts an ϵ-greedy strategy, selecting a random action with probability ϵ, while the optimal action, $\arg\max_a Q(s, a)$, is selected with probability $1-\epsilon$.

4 Experimental Evaluation and Results

This section provides a detailed analysis of the data used and the results produced by the model we implemented.

4.1 Data Preprocessing

The models in our system use important parameters such as "Downlink traffic," "Duration," and "Latency" as input features, with the "Offloading Decision" as the target variable. These features are crucial, as they provide valuable information about network condition and device performance.

- The "Start Time" and "End Time" columns are transformed into date time format to simplify processing.
- By incorporating a new column, "Offloading Decision," which indicates if a task will be offloaded to the edge cloud or processed locally on a mobile device. If download traffic is greater than upload traffic, the task is offloaded to the edge cloud; if not they are processed on the mobile device.
- By introducing a new column, "Latency," which computes latency based on the offloading decision. In the case of task offloading to the edge cloud, an additional 5 s is included in the task duration. If not, the task's duration used directly.
- By adding a new "Throughput" column, which determines the throughput of each task by dividing the total traffic (uploads + downloads) by the related latency.
- The edge cloud usage percentage is obtained by partitioning the total latency of tasks offloaded to the cloud by the total latency of all tasks, and then multiplying the result by 100.

4.2 Experimental Results and Discussion

Figure 2 demonstrates the performance differences between the four classifiers (RF, SVM, GB, and DQN) regarding precision for the two classes: Edge Cloud and Mobile.

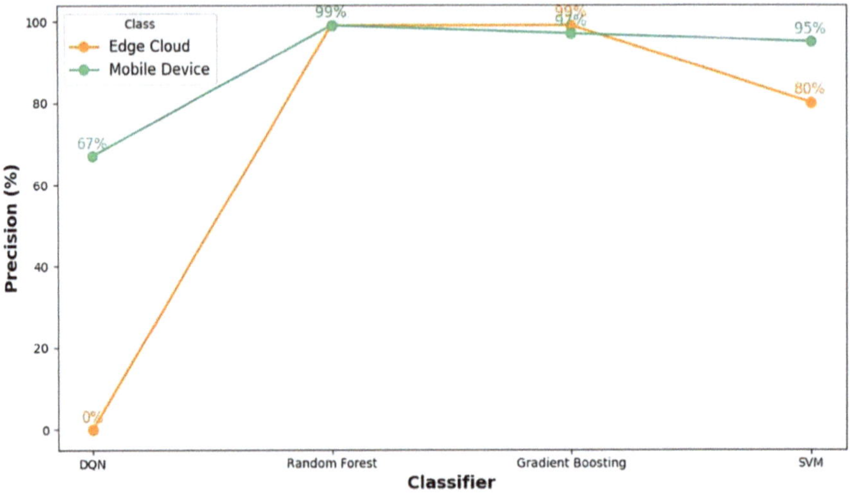

Fig. 2 Comparison of precision across classifiers

Device. In the edge cloud class, precision begins at 0% with DQN, reflecting a total lack of correct predictions. However, a remarkable improvement is seen with Random Forest and Gradient Boosting, both achieving an optimal precision of 99%. With SVM, precision falls to 80%, indicating a decline in performance for this class. For the mobile device class, DQN attains a precision of 67%, which is notably lower than that of the other models. RF and GB achieved outstanding precision of 99%, highlighting their effectiveness in handling this class. SVM, while slightly lower, still maintains a high precision of 95%. In general, Random Forest and Gradient Boosting excel with consistently strong performance across both classes, whereas DQN demonstrates significant limitations, particularly for the edge cloud class.

The effectiveness of the four models is analyzed based on accuracy. As represented in Fig. 3, the RF model excels with optimal results for both offloading classes. Gradient Boosting, while slightly behind, performs nearly on par with RF. The SVM model, however, yields less favorable results, especially due to its underperformance in the edge cloud class. Conversely, the DQN model demonstrates relatively low accuracy, implying that it needs further refinement to better suit this specific task.

Figure 4 shows the recall performance of the DQN, RF, GB, and SVM classifiers across the edge cloud and mobile device classes. The DQN model records a 0% recall for the edge cloud class, underscoring its failure to recognize any samples from this category. Conversely, Random Forest and Gradient Boosting demonstrate optimal performance, each attaining a perfect recall of 100%, whereas SVM achieves a slightly lower yet still strong recall of 99%. For the mobile device class, the DQN model again fails completely, registering a recall of 0%. With a perfect recall of 100%, Random Forest leads in performance, closely followed by Gradient Boosting at 99%, indicating similarly strong results. In contrast, the SVM model exhibits a significant decline, with a recall of only 53%, which suggests difficulty in recognizing this class.

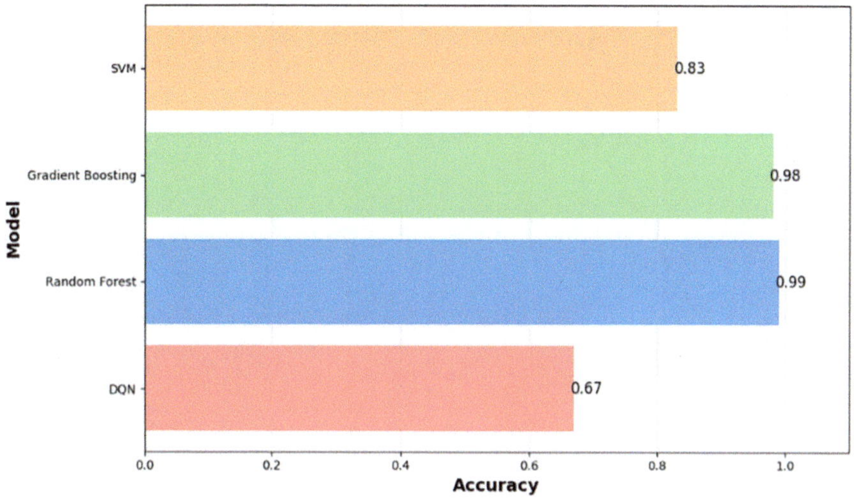

Fig. 3 Model accuracy comparison

In general, RF and GB show strong and consistent performance across both classes, whereas SVM produces variable results, and DQN is unable to identify either class.

Figure 5 illustrates a comparison of the F1-scores achieved by the Random Forest, Gradient Boosting, SVM, and DQN classifiers for the edge cloud and mobile device classes. DQN scores 0% for the edge cloud class, reflecting a total inability to correctly predict instances from this category. On the other hand, Random Forest delivered an outstanding F1-score of 100%, indicating perfect classification. Gradient

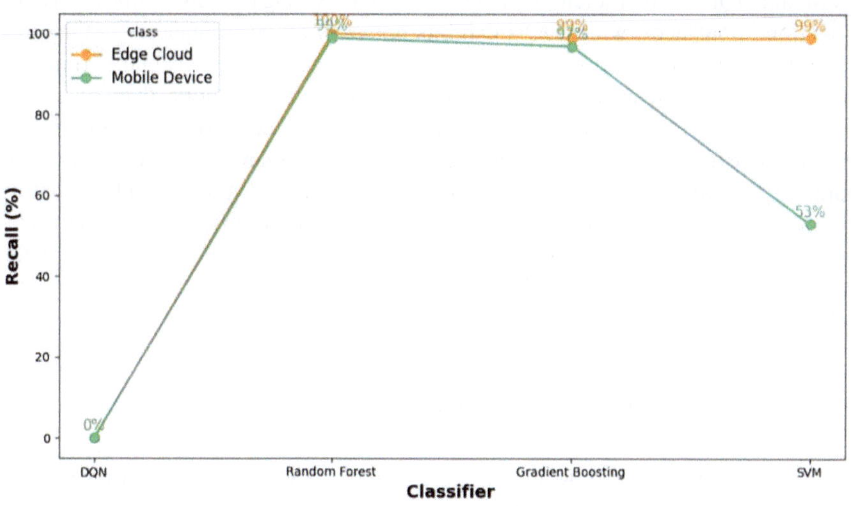

Fig. 4 Comparison of recall across classifiers

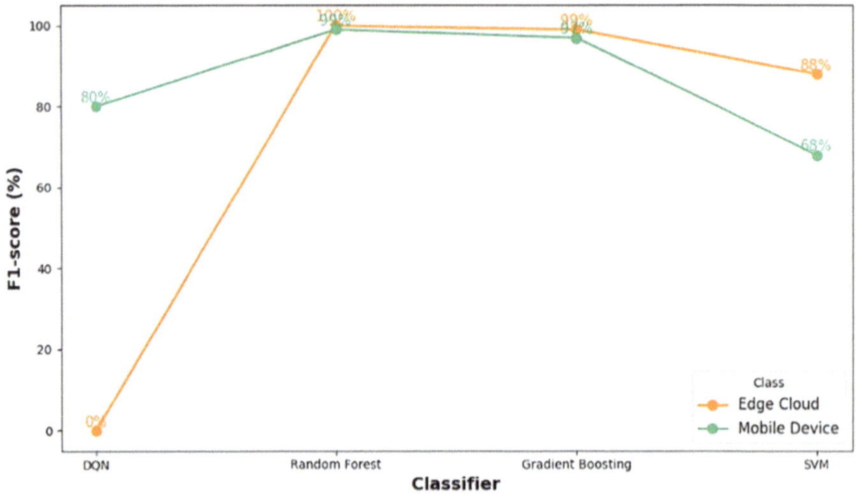

Fig. 5 Comparison of F1-score across classifiers

Boosting performed nearly as well with a score of 99%, whereas SVM, though lower, still performs reasonably well with an F1-score of 88%. In mobile device class, DQN achieved a moderate F1-score of 80%, while Random Forest and Gradient Boosting excelled, scoring 100% and 99%, respectively. Meanwhile, SVM saw a considerable decrease, with its F1-score falling to 68%. The results emphasize the dominance of Random Forest and Gradient Boosting across both classes, while SVM demonstrates inconsistent performance and DQN proves ineffective for the edge cloud class.

Although the models used show satisfactory results, certain drawbacks remain. SVM and DQN are inefficient, with SVM affected by long execution times and classification imbalances, while DQN exhibits poor overall performance. On the other hand, RF and Gradient Boosting provide better accuracy and acceptable runtimes but may face challenges in handling more complex or dynamic data scenarios. Furthermore, the impact of variations in latency and network traffic on system performance have not been explored in-depth. Future research could focus on incorporating lightweight neural networks designed for MEC systems and evaluating the models in real, dynamic environments. What's more, investigating the trade-offs between performance, complexity, and energy efficiency could enhance the approach's applicability in Mobile Edge Computing (MEC).

5 Conclusion and Future Work

We created a novel IDS for Mobile Edge Computing (MEC), incorporating ML and neural network techniques including RF, SVM, GB, and Deep Q-Network (DQN). After evaluating the system with the Guangzhou and Shanghai datasets, we found that it successfully supports task offloading decisions in MEC, further validating the strength of our approach. Future research will focus on incorporating lightweight neural networks tailored for MEC, testing the system in dynamic real-world environments to further refine it. Additionally, examining the trade-offs between performance, complexity, and energy efficiency will be crucial to enhance the approach's applicability to MEC.

Acknowledgements We sincerely appreciate the support of the CNRST for their backing through the PASS (Ph.D.-Associate Scholarship) program.

References

1. R.-F. Liao, H. Wen, J. Wu, F. Pan, A. Xu, and H. Song, "Security Enhancement for Mobile Edge Computing Through Physical Layer Authentication," IEEE Access, vol. 7, pp. 116390–116401, 2019, https://doi.org/10.1109/ACCESS.2019.2934122.
2. B. Ali, M. A. Gregory, and S. Li, "Multi-Access Edge Computing Architecture, Data Security and Privacy: A Review," IEEE Access, vol. 9, pp. 18706–18721, 2021, https://doi.org/10.1109/ACCESS.2021.3053233.
3. H. Attou, A. Guezzaz, S. Benkirane, and M. Azrour, "A New Secure Model for Cloud Environments Using RBFNN and AdaBoost," SN Comput. Sci., vol. 6, no. 2, p. 188, Feb. 2025, https://doi.org/10.1007/s42979-025-03691-1.
4. H. Assmi, A. Guezzaz, S. Benkirane, M. Azrour, S. Jabbour, N. Innab, A. Alabdulatif, "A Robust Security Detection Strategy for Next Generation IoT Networks. | EBSCOhost." Accessed: May 19, 2025. [Online]. Available: https://openurl.ebsco.com/contentitem/10.32604%2Fcmc.2024.059047?sid=ebsco:plink:crawler&id=ebsco:10.32604%2Fcmc.2024.059047
5. S. Dargaoui, M. Azrour, A. E. Allaoui, A. Guezzaz, and S. Benkirane, "Machine learning for security boosting in Internet of Things environments," in Recent Advances in Internet of Things Security, CRC Press, 2025.
6. A. Blali, Dargaoui,Souhayla, Azrour,Mourade, Guezzaz,Azidine, Amounas,Fatima, and A. and Alabdulatif, "Analysis of deep learning-based intrusion detection systems in IoT environments," EDPACS, vol. 0, no. 0, pp. 1–35, https://doi.org/10.1080/07366981.2025.2498222.
7. S. Dargaoui, M. Azrour, A. El Allaoui, A. Guezzaz, A. Alabdulatif, and S. Ahmad, "ECC-Based Anonymous and Multi-factor Authentication Scheme for IoT Environment. | EBSCOhost." Accessed: May 19, 2025. [Online]. Available: https://openurl.ebsco.com/contentitem/10.3991%2Fijoe.v21i01.51633?sid=ebsco:plink:crawler&id=ebsco:10.3991%2Fijoe.v21i01.51633
8. M. Mohy-Eddine, A. Guezzaz, S. Benkirane, M. Azrour, and K. Bella, "A monitoring system with deep learning for IoT smart environments security," in Recent Advances in Internet of Things Security, CRC Press, 2025.
9. G. Wu, X. Chen, Z. Gao, H. Zhang, S. Yu, and S. Shen, "Privacy-preserving offloading scheme in multi-access mobile edge computing based on MADRL," J. Parallel Distrib. Comput., vol. 183, p. 104775, Jan. 2024, https://doi.org/10.1016/j.jpdc.2023.104775.
10. S. Amaouche, A. Guezzaz, S. Benkirane, and M. Azrour, "A Robust Model for Predicting Abnormal Behavior in Vehicular Networks using AdaBoost and Chi-Square," Wirel. Pers. Commun., vol. 138, no. 4, pp. 2583–2611, Oct. 2024, https://doi.org/10.1007/s11277-024-11615-0.
11. X. Liu, J. Liu, and W. Li, "Truthful mechanism for joint resource allocation and task offloading in mobile edge computing," Comput. Netw., vol. 254, p. 110796, Dec. 2024, https://doi.org/10.1016/j.comnet.2024.110796.

12. C. Hazman, A. Guezzaz, S. Benkirane, and M. Azrour, "A smart model integrating LSTM and XGBoost for improving IoT-enabled smart cities security," Clust. Comput., vol. 28, no. 1, p. 70, Nov. 2024, https://doi.org/10.1007/s10586-024-04780-1.
13. K. Bella, A. Guezzaz, S. Benkirane, M. Azrour, Y. Fouad, MS. Benyeogor, and N. Innab, "An efficient intrusion detection system for IoT security using CNN decision forest," PeerJ Comput. Sci., vol. 10, p. e2290, Sep. 2024, https://doi.org/10.7717/peerj-cs.2290.
14. M. Mohy-eddine, A. Guezzaz, S. Benkirane, and M. Azrour, "Malicious detection model with artificial neural network in IoT-based smart farming security," Clust. Comput., vol. 27, no. 6, pp. 7307–7322, Sep. 2024, https://doi.org/10.1007/s10586-024-04334-5.
15. S. Dargaoui et al., "Internet-of-Things-Enabled Smart Agriculture: Security Enhancement Approaches," in 2024 4th International Conference on Innovative Research in Applied Science, Engineering and Technology (IRASET), May 2024, pp. 1–5. https://doi.org/10.1109/IRASET60544.2024.10548705.
16. T. Anwar, G. A. Khan, Z. Ashraf, Z. A. Ansari, R. Ahmed, and M. Azrour, "The Combination of Blockchain and the Internet of Things (IoT): Applications, Opportunities, and Challenges for Industry," in Blockchain and Machine Learning for IoT Security, Chapman and Hall/CRC, 2024.
17. C. Wang, Z. Yuan, P. Zhou, Z. Xu, R. Li, and D. O. Wu, "The Security and Privacy of Mobile-Edge Computing: An Artificial Intelligence Perspective," IEEE Internet Things J., vol. 10, no. 24, pp. 22008–22032, Dec. 2023, https://doi.org/10.1109/JIOT.2023.3304318.
18. J. A. Alzubi, O. A. Alzubi, A. Singh, and T. Mahmod Alzubi, "A blockchain-enabled security management framework for mobile edge computing," Int. J. Netw. Manag., vol. 33, no. 5, p. e2240, 2023, https://doi.org/10.1002/nem.2240.
19. Z. Ning, H. Hu, X; Wang, L. Guo, G. Wang, and X. Goa, "Mobile Edge Computing and Machine Learning in the Internet of Unmanned Aerial Vehicles: A Survey," ACM Comput Surv, vol. 56, no. 1, p. 13:1–13:31, août 2023, https://doi.org/10.1145/3604933.
20. L. Zhang, S. Lai, J. Xia, C. Gao, D. Fan, and J. Ou, "Deep reinforcement learning based IRS-assisted mobile edge computing under physical-layer security," Phys. Commun., vol. 55, p. 101896, Dec. 2022, https://doi.org/10.1016/j.phycom.2022.101896.
21. A. Singh, S. C. Satapathy, A. Roy, and A. Gutub, "AI-Based Mobile Edge Computing for IoT: Applications, Challenges, and Future Scope," Arab. J. Sci. Eng., vol. 47, no. 8, pp. 9801–9831, Aug. 2022, https://doi.org/10.1007/s13369-021-06348-2.
22. Y. Li, Q. Cheng, X. Liu, and X. Li, "A Secure Anonymous Identity-Based Scheme in New Authentication Architecture for Mobile Edge Computing," IEEE Syst. J., vol. 15, no. 1, pp. 935–946, Mar. 2021, https://doi.org/10.1109/JSYST.2020.2979006.
23. A. Rahman, G. Wu, and A. M. Liton, "Mobile edge computing for internet of things (IoT): security and privacy issues," Indones. J. Electr. Eng. Comput. Sci., vol. 18, no. 3, p. 1486, Jun. 2020, https://doi.org/10.11591/ijeecs.v18.i3.pp1486-1493.
24. K. Cao, Y. Liu, G. Meng, and Q. Sun, "An Overview on Edge Computing Research," IEEE Access, vol. 8, pp. 85714–85728, 2020, https://doi.org/10.1109/ACCESS.2020.2991734.
25. B. Huang, Z. Li, P. Tang, S. Wang, J. Zhao, H. Hu, W. Li, and V. Chang, "Security modeling and efficient computation offloading for service workflow in mobile edge computing," Future Gener. Comput. Syst., vol. 97, pp. 755–774, Aug. 2019, https://doi.org/10.1016/j.future.2019.03.011.
26. Y. Xiao, Y. Jia, C. Liu, X. Cheng, J. Yu, and W. Lv, "Edge Computing Security: State of the Art and Challenges," Proc. IEEE, vol. 107, no. 8, pp. 1608–1631, Aug. 2019, https://doi.org/10.1109/JPROC.2019.2918437.
27. C. Zhang and Z. Zheng, "Task migration for mobile edge computing using deep reinforcement learning," Future Gener. Comput. Syst., vol. 96, pp. 111–118, Jul. 2019, https://doi.org/10.1016/j.future.2019.01.059.
28. R. Roman, J. Lopez, and M. Mambo, "Mobile edge computing, Fog et al.: A survey and analysis of security threats and challenges," Future Gener. Comput. Syst., vol. 78, pp. 680–698, Jan. 2018, https://doi.org/10.1016/j.future.2016.11.009.

29. D. He, S. Chan, and M. Guizani, "Security in the Internet of Things Supported by Mobile Edge Computing," IEEE Commun. Mag., vol. 56, no. 8, pp. 56–61, Aug. 2018, https://doi.org/10.1109/MCOM.2018.1701132.
30. H. Zhao, W. Du, W. Liu, T. Lei, and Q. Lei, "QoE Aware and Cell Capacity Enhanced Computation Offloading for Multi-Server Mobile Edge Computing Systems with Energy Harvesting Devices," in 2018 IEEE SmartWorld, Ubiquitous Intelligence & Computing, Advanced & Trusted Computing, Scalable Computing & Communications, Cloud & Big Data Computing, Internet of People and Smart City Innovation (SmartWorld/SCALCOM/UIC/ATC/CBDCom/IOP/SCI), Oct. 2018, pp. 671–678. https://doi.org/10.1109/SmartWorld.2018.00133.
31. S. Wu, W. Xia, W. Cui, Q. Chao, Z. Lan, F. Yan, and L. Shen, "An Efficient Offloading Algorithm Based on Support Vector Machine for Mobile Edge Computing in Vehicular Networks," in 2018 10th International Conference on Wireless Communications and Signal Processing (WCSP), Oct. 2018, pp. 1–6. https://doi.org/10.1109/WCSP.2018.8555695.

AI-Driven Security Automation in DevOps: Cybersecurity in CI/CD Pipelines

Aya El Mazhari, Ayoub Bahnasse, Mouaad Mohy-Eddine, and Mourad Zegrari

Abstract The rapid rise of DevOps has substantially expedited software development and deployment processes. However, this shift has also heightened security vulnerabilities, making DevOps pipelines increasingly vulnerable to cyber threats. The integration of Artificial Intelligence (AI) can strengthen the security of DevOps pipelines through advanced threat identification and mitigation strategies. AI is now being deployed within DevOps to automate repetitive processes, reduce manual intervention, and enhance Continuous Integration and Continuous Deployment (CI/CD). However, organizations struggle to integrate AI-driven security, resulting in fragmented approaches that are inadequate for today's evolving threat landscape. This paper offers practical guidance for seamlessly integrating AI into DevOps, enhancing security, trust, and efficiency. We examine how AI can be applied within CI/CD pipelines, address key challenges, and explore the future of AI-enhanced cybersecurity in DevOps. Our goal is to empower organizations to leverage AI for a robust and secure DevOps lifecycle.

Keywords DevOps · DevSecOps · AI · CI/CD · ML · Security automation

1 Introduction

In our ever-connected world, cybersecurity threats remain a constant concern for organizations. Cybercriminals are becoming smarter, utilizing advanced techniques to exploit weaknesses, placing sensitive data at risk, and disrupting operations. With the expansion of software-based services, the potential for attacks has increased, making robust security measures more crucial than ever.[1]

---

[1] https://owasp.org/www-project-top-10-ci-cd-security-risks/

A. El Mazhari (✉) · A. Bahnasse · M. Mohy-Eddine · M. Zegrari
Delta Lab, Hassan II University, Casablanca, Morocco
e-mail: ayaelmazhari@gmail.com

M. Azrour and A. Alabdulatif (eds.), *AI-Driven Security for Next-Generation IoT Systems*, https://doi.org/10.1007/978-3-032-08784-3_6

The shift toward DevOps [1] practices brings its own set of security challenges. The focus on automation and rapid deployment in Continuous Integration and Continuous Deployment (CI/CD) [2] pipelines can lead to vulnerabilities if security isn't prioritized from the start. Traditional security methods often struggle to keep up with the fast pace of DevOps, which has given rise to the concept of DevSecOps [3]; This approach integrates security practices throughout the entire DevOps lifecycle [4].

Within the DevSecOps framework, AI-driven security automation has emerged as a transformative solution [5]. Artificial Intelligence (AI) and Machine Learning (ML) can analyze massive amounts of data, identify patterns that indicate malicious activity, and automate responses to threats in real time. This proactive security strategy allows organizations to stay a step ahead of evolving threats while sustaining the speed and flexibility that DevOps demands [6].

This paper delves into how AI-driven security automation can address these challenges and boost security, trust, and efficiency in DevOps. By exploring the applications, advantages, and obstacles of incorporating AI into the DevSecOps lifecycle, this research aims to equip organizations with the knowledge to confidently harness AI for a secure and resilient DevOps environment.

The rest of the paper is structured as follows, in the second section, we investigate the applications and use cases of AI for security automation, as well as key AI techniques employed. We also provide a comparative analysis of Traditional and AI-enabled security approaches. In the third section, we present the benefits and challenges of AI-driven security automation, along with strategies to mitigate these challenges. In the fourth section, we discuss the future directions of AI-powered cybersecurity. Finally, we present a conclusion for our work.

2 AI in Cybersecurity for DevOps: Current Trends

2.1 DevOps, DevSecOps, and CI/CD Pipeline

DevOps, an agile approach integrating software development and IT operations, has seen widespread adoption in recent years [7]. This methodology emphasizes collaboration, automation, and continuous improvement, enabling faster and more efficient software delivery [8]. However, the accelerated DevOps pace can introduce security vulnerabilities if not addressed proactively.

The emergence of DevSecOps integrates security practices throughout the DevOps lifecycle, see Fig. 1. DevSecOps embedding security considerations at every software development stage, from planning to deployment.

The CI/CD pipeline, a core DevOps component, is vulnerable to security threats. CI/CD pipelines automate build, test, and deploy as illustrated in Fig. 2, but can also be susceptible to attacks compromising software integrity or infrastructure.

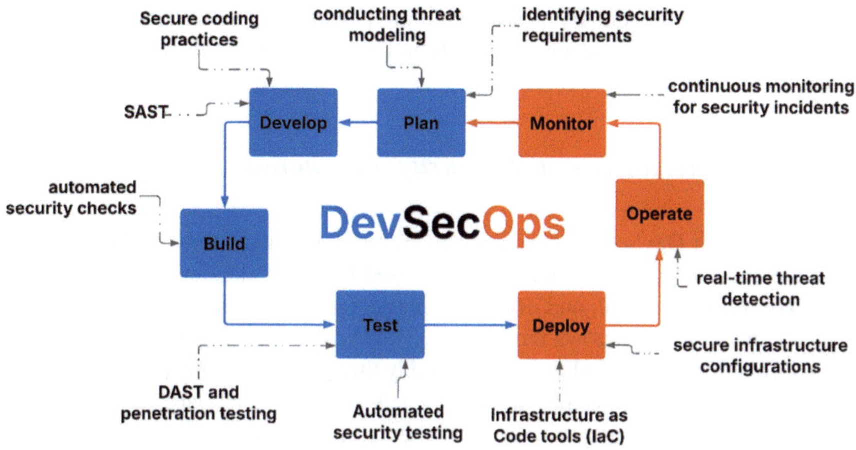

Fig. 1 DevSecOps implementation

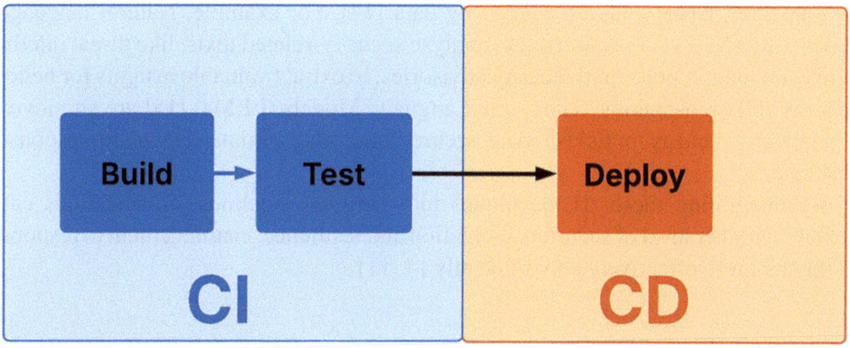

Fig. 2 CI/CD pipeline

2.2 *AI Applications in DevSecOps*

AI is used to automatically analyze source code to identify vulnerabilities, like buffer overflows, SQL injection, or cross-site scripting [9]. AI-powered static and dynamic code analysis tools can detect vulnerabilities in an early stage of the development process and automatically test running applications for vulnerabilities by simulating real-world attacks [6]. The AI-powered tools can automatically generate and apply patches, reducing the exposure window and the burden on security teams [10]. Similarly, AI models can be trained to analyze network traffic, user behavior, and system logs and then identify potential anomalies that may cause a security breach. Moreover, our AI models can identify anomalies and patterns revealing malicious activity in real time when trained on large datasets of cyber threats, allowing for rapid response and containment [11]. Finally, AI can continuously monitor and assess the evolving risk landscape, adjusting security controls and policies for any emerging threats.

2.3 Key AI Techniques for Security Automation

The use of AI in DevSecOps leverages a variety of ML algorithms and techniques. First, there is supervised learning, which is used for tasks like detecting vulnerabilities, classifying malware, and detecting intrusions. Where the model is trained on labeled data to identify known threats [12]. Then unsupervised learning is employed for tasks like identifying unknown or zero-day threats, detecting insider threats, and uncovering hidden relationships in complex network data. Afterward, Reinforcement Learning helps create adaptive security systems that learn from their interactions with the environment, constantly enhancing their ability to detect and respond to threats [13]. Deep learning (DL) algorithms are particularly effective at detecting complex attack patterns, including advanced persistent threats and insider threats, by analyzing large volumes of security data [14]. For example, Natural language Processing (NLP) [12] is utilized to analyze security-related texts, like threat intelligence, incident reports, and security advisories, to extract valuable insights for better security decision-making. The Large Language Models (LLMs) [15] are employed to generate security policies, write secure code, and automate incident response procedures.

By integrating these AI techniques into DevOps pipelines, organizations can achieve a higher level of security automation and resilience, enabling them to respond to threats more effectively and efficiently [4, 11].

2.4 AI-Driven Security Automation Platforms and Approaches

While there isn't a single universally adopted "framework" for AI-driven security automation, several key platforms, libraries, and approaches are commonly used. The choice often depends on the specific security needs and the existing infrastructure of an organization. Many Security Information and Event Management (SIEM) platforms like Splunk, IBM QRadar, and LogRhythm now incorporate AI and ML capabilities. These platforms ingest security logs from various sources, and the AI components analyze this data to detect anomalies, prioritize alerts, and even automate some response actions. Ovabor et al. [12] mentions IBM QRadar Advisor with Watson and Microsoft Azure Sentinel as examples of such a platform. Moreover, security orchestration, automation, and response (SOAR) tools like Palo Alto Networks Cortex XSOAR and Splunk Phantom play a crucial role in automating security workflows. While not exclusively AI-driven, they often integrate with AI/

ML platforms to enrich threat intelligence and automate incident response procedures. Redding [6] discusses incorporating AI within CI/CD pipelines for automated responses, which aligns with the functionality of SOAR tools.

Cloud providers like AWS, Azure, and GCP offer security services with built-in AI for threat detection and mitigation. These services leverage the vast data and computing resources of the cloud to provide scalable and efficient security automation. Abiona et al. [4] mentions the use of AI/ML in cloud security tools. Also, several open-source libraries and frameworks like TensorFlow, PyTorch, and scikit-learn can be used to develop custom AI models for security automation. These frameworks offer flexibility and customization but require expertise in ML and cybersecurity. Finally, Threat Intelligence Platforms (TIPs), such as Recorded Future and ThreatConnect, aggregate and analyze threat data from various sources. AI plays a crucial role in enriching this data, identifying emerging threats, and providing actionable insights. Ovabor et al. [12] discusses AI-driven threat intelligence in detail.

Integrating security practices into the DevOps pipeline is becoming increasingly common, and AI plays a vital role in automating security testing and vulnerability remediation within this framework. Redding [6] emphasizes the importance of AI in DevSecOps for enhanced security. Besides AI is essential for enforcing Zero Trust principles by continuously verifying user identities and access privileges based on real-time risk assessments.

The effectiveness of AI-driven security automation heavily depends on the quality of training data and the ability to effectively integrate diverse data sources and understanding how AI models arrive at their decisions is crucial for building trust and improving their accuracy.

Integrating security into the DevOps pipeline through DevSecOps is increasingly vital, with AI playing a crucial role in automating security testing and vulnerability remediation [6]. AI's capabilities are also essential for enforcing Zero Trust principles by continuously verifying user identities and access privileges based on real-time risk assessments [16]. The effectiveness of AI-driven security automation heavily depends on the quality of training data and the ability to effectively integrate diverse data sources within the Model Context Protocol. Furthermore, understanding how AI models arrive at their decisions is crucial for building trust and improving their accuracy, highlighting the importance of explainability and transparency in AI-driven security practices [17]. By addressing these key considerations, organizations can leverage AI to enhance security, streamline processes, and foster a more resilient and efficient DevSecOps environment [6].

By combining these platforms, tools, and approaches strategically, organizations can effectively automate many aspects of security monitoring and incident response. However, it's essential to carefully evaluate specific needs and choose solutions that integrate well with existing systems and processes.

2.5 Comparing Traditional and AI-Enabled Security

While traditional security measures remain important, AI offers significant advantages as shown in Table 1. It shifts security from reactive to proactive, while traditional security primarily reacts to known threats. AI-driven security can proactively identify potential threats and vulnerabilities before they are exploited. Samarakoon et al. [18] mentions AI's shift "from detecting past-known threats to predicting new ones."

AI-driven security identifies potential threats before exploitation. Instead of relying on signatures of known malware, AI uses anomaly detection to identify zero-day attacks and insider threats that traditional methods often miss. Redding [6] highlights AI's strength in "detecting anomalies" and identifying "attack patterns that may go unnoticed by conventional security tools."

Furthermore, AI automates threat detection and response, making it much faster and more scalable than traditional methods, especially important with the increasing volume and velocity of cyberattacks. Samarakoon et al. [18] discusses how AI enhances "the speed and precision of threat detection." Finally, AI/ML models learn and adapt to evolving threat landscapes. Traditional systems require manual updates to stay current. Samarakoon et al. [18] underscores how "conventional reactive security approaches [...] are insufficient against intelligent attacks," making "self-adaptive intelligent security systems" necessary.

Table 1 Comparison of traditional and AI-driven security measures

Feature	Traditional security	AI-driven security
Approach	Reactive	Proactive
Threat detection	Signature-based, rule-based	Anomaly detection, behavioral analysis
Speed	Slower, often manual	Faster, automated
Accuracy	Can be less accurate due to reliance on known signatures	Higher accuracy due to anomalies and patterns identification
Scalability	Difficult to scale manually	Easily scalable due to automation
Adaptation	Requires manual updates to rules and signatures	Adapts to new threats and evolving patterns automatically
Cost	Less expensive initially but requires significant ongoing human resources	Expensive to implement initially but requires less human intervention over time
Effectiveness	Effective against known threats	Effective against known and unknown (zero-day) threats
Examples	Firewalls, intrusion detection systems, antivirus software	ML-based intrusion detection, user and entity behavior analytics, AI-powered threat intelligence

3 AI-Powered Cybersecurity for Software Development: Benefits, Challenges, and Mitigations

AI is transforming cybersecurity in software development, offering significant advantages in threat detection, vulnerability management, and incident response. AI algorithms master analyzing vast datasets and identifying patterns indicative of malicious activity such as malware, ransomware, and unauthorized access [12].

However, several challenges need to be addressed. These include ensuring data quality and mitigating bias in AI models [6], improving the explainability and transparency of AI-driven decisions, and developing scalable and efficient AI solutions. Finally, securing AI systems themselves against attacks is crucial [7].

3.1 Advantages of AI-Driven Security Automation in DevOps Pipelines

This research explored the integration of AI-driven security automation within DevOps pipelines. The analysis revealed that AI offers substantial advantages, enhancing cybersecurity while preserving the speed and agility inherent in DevOps practices. AI-powered tools demonstrated enhanced real-time threat detection and response, exhibiting higher accuracy and lower false positive rates compared to traditional security solutions. This proactive approach minimizes successful cyberattack risks, safeguarding pipeline integrity and infrastructure. Specifically, [6] found that AI-driven security frameworks reduced the time required for security tasks by 30%. Also, AI automation significantly reduces time and effort for security tasks such as vulnerability scanning and incident response. This efficiency allows for maintaining rapid software delivery while ensuring robust security measures. Battina [7] illustrated an improvement of 93.75% in average detection time, a reduction of 75% in False Positive rate, a reduction of 95.83% in threat response time, and a reduction of 70% in the number of undetected attacks after the integration of AI/ML in cybersecurity. Additionally, AI automation supports continuous compliance with industry regulations. This capability simplifies adherence to complex security standards [19].

AI-driven security automation effectively scales with complex DevOps environments, enabling visibility and monitoring across the entire pipeline. And AI systems continuously learn and evolve, improving threat detection and mitigation. This adaptability ensures a robust security posture aligned with the changing threat landscape.

3.2 Ethical and Regulatory Concerns and Risks of Over-Reliance on AI-Driven Security

The integration of AI-driven security tools into existing infrastructure is challenging, as legacy systems may not accommodate new technologies seamlessly. Compatibility issues from legacy applications, databases, and protocols require extensive integration and migration. Additionally, training effective AI models demand large, diverse datasets, which may not be readily available, especially for emerging threats. The complexity of feature engineering, hyperparameter tuning, and model validation can further complicate deploying AI-powered security solutions. Several sources such as [20] and [21] also emphasize the need for future research in specific areas, such as efficient, scalable AI algorithms, seamless DevSecOps integration, and AI model security itself, to fully realize AI's potential in this domain, as shown in Table 2.

By tackling these challenges head-on, we can unlock the true power of AI to build a safer future for software development.

Table 2 Overview of the key concerns related to AI-driven security, their definitions, and mitigation strategies

Concern	Definition	Mitigation strategy
Bias and discrimination	Training on biased data can lead to discriminatory outcomes	Use diverse training data, employ explainable AI to detect bias, and monitor model outputs
Lack of transparency	The "black box" nature of AI models hinders understanding and trust	Develop interpretable models and use XAI for insights. Document AI usage in security operations
Over-reliance	Automating security tasks can lead to over-dependence on AI and a decline in human expertise	Balance automation and human oversight. Train human operators to critically evaluate AI, continuous training for security professionals
Privacy concerns	Data collection risks breaches and misuse	Adhering to regulations, transparency, and privacy-preserving techniques
Security of AI systems	AI models are vulnerable to attacks	Develop robust models and security, defend against attacks, regular testing
Regulatory and compliance	Evolving laws pose challenges	Stay informed, and implement data governance, risk assessments, and industry best practices

4 Future Outlook on AI-Powered Cybersecurity in DevOps

The future of Artificial Intelligence (AI) in cybersecurity for software development holds significant promise, with anticipated advancements in automation, proactive defense mechanisms, and enhanced security across the software development lifecycle. AI-powered solutions are expected to play a pivotal role in autonomously identifying and remediating code vulnerabilities, thereby reducing the need for manual intervention [20]. Moreover, AI will increasingly automate a variety of security testing processes, including penetration testing and fuzzing, resulting in more efficient and comprehensive assessments of software robustness [4]. In the domain of incident response, AI is poised to streamline detection, containment, and recovery workflows, significantly accelerating the overall response time to security breaches [6].

AI's capacity to process and analyze vast quantities of data will further enable the early identification of emerging threats and the prediction of potential attack vectors, facilitating proactive defense strategies. This capability will also support the detection of previously unknown threats and vulnerabilities, even in the absence of recognizable attack signatures. Furthermore, AI-driven security systems will evolve to autonomously adapt to dynamic threat environments and draw from historical incidents to enhance their defensive efficacy over time [22].

The continued development of AI-driven threat intelligence will refine and expand the capabilities of AI-powered code review, thereby maximizing the potential of AI in DevOps and building secure, resilient, and high-quality software [17]. As organizations adopt DevSecOps practices, AI will become indispensable in ensuring an adaptive and efficient strategy for mitigating cyber threats in CI/CD pipelines [23]. Ethical considerations are vital in the deployment of AI in cybersecurity [11].

Explainable AI will gain prominence as organizations seek to understand and trust AI-driven security decisions, providing insights into the reasoning behind AI's actions and enabling better human oversight [17]. Privacy-preserving techniques such as federated learning will also gain importance, allowing AI models to be trained on decentralized data sources while protecting sensitive information [23]. The focus will be on promoting transparency, accountability, and fairness in AI systems, thereby building trust and mitigating potential biases [23]. Addressing bias, ensuring data privacy, and navigating the delicate balance between automation and job displacement are critical challenges [24].

As AI becomes more deeply integrated into cybersecurity, ethical frameworks and governance structures will be essential to ensure its responsible and beneficial use [12]. Ongoing research should focus on developing resilient AI models capable of adapting to rapidly changing threat landscapes and addressing the ethical implications of increasingly autonomous security systems [22].

5 Conclusion

This study has investigated the transformative impact of Artificial Intelligence (AI)-driven security automation within the context of DevOps. Specifically, it analyzed the integration of AI across various phases of the Continuous Integration and Continuous Deployment (CI/CD) pipeline. The deployment of AI technologies in domains such as malware detection, intrusion prevention, and SIEM systems demonstrates considerable benefits in terms of operational speed, detection accuracy, and the establishment of a proactive security posture. Nonetheless, the integration of AI into DevSecOps is accompanied by several critical challenges. These include the intricacies of developing AI models that maintain both high accuracy and fairness, the necessity for meticulous data management, and the increasing threat of adversarial attacks targeting AI-based systems. Furthermore, the incorporation of AI into legacy infrastructures presents non-trivial technical and financial hurdles, often necessitating substantial adaptation and investment. Overreliance on AI solutions, in the absence of adequate human oversight, also introduces significant ethical concerns regarding the governance and accountability of security decisions.

Despite these limitations, the trajectory for AI in DevSecOps remains promising. Future progress will depend on advancing robust and interpretable AI models, alongside the development of standardized implementation and governance frameworks. Addressing these challenges holistically will enable organizations to fully harness the benefits of AI-powered security automation, thereby enhancing the efficiency, security, and resilience of modern software development lifecycles. Moreover, the advancement of explainable AI will be crucial in promoting transparency and trust in automated security decision-making processes. Finally, fostering interdisciplinary collaboration among cybersecurity professionals, AI researchers, and DevOps practitioners will be essential to realizing a cohesive and secure software development ecosystem.

References

1. Nasir Qureshi, J., Shoaib Farooq, M., Ali, U., Khelifi, A., Atal, Z.: Exploring the Integration of Blockchain and Distributed DevOps for Secure, Transparent, and Traceable Software Development. IEEE Access. 13, 15489–15502 (2025). https://doi.org/10.1109/ACCESS.2024.3509036.
2. Clark, T.: Introduction to Continuous Integration and Continuous Deployment. In: CI/CD Unleashed. pp. 1–18. Apress, Berkeley, CA (2025). https://doi.org/10.1007/979-8-8688-1209-5_1.
3. Prates, L., Pereira, R.: DevSecOps practices and tools. Int. J. Inf. Secur. 24, 11 (2025). https://doi.org/10.1007/s10207-024-00914-z.
4. Oluwatosin Oluwatimileyin Abiona, Oluwatayo Jacob Oladapo, Oluwole Temidayo Modupe, Oyekunle Claudius Oyeniran, Adebunmi Okechukwu Adewusi, Abiola Moshood Komolafe: The emergence and importance of DevSecOps: Integrating and reviewing security practices within the DevOps pipeline. World J. Adv. Eng. Technol. Sci. 11, 127–133 (2024). https://doi.org/10.30574/wjaets.2024.11.2.0093.

5. Fu, M., Pasuksmit, J., Tantithamthavorn, C.: AI for DevSecOps: A Landscape and Future Opportunities. ACM Trans. Softw. Eng. Methodol. 34, 1–61 (2025). https://doi.org/10.1145/3712190.
6. Redding, P.: Securing DevOps Pipelines with AI-Powered Threat Detection and Mitigation, (2025).
7. Battina, D.S.: Ai and devops in information technology and its future in the united states. IJCRT. 2320--2882 (2021).
8. Alenezi, M., Zarour, M., Akour, M.: Can Artificial Intelligence Transform DevOps?, https://arxiv.org/abs/2206.00225, (2022). https://doi.org/10.48550/ARXIV.2206.00225.
9. Oloyede, J.: Leveraging Artificial Intelligence for Advanced Cybersecurity Threat Detection and Prevention. SSRN Journal. (2024). https://doi.org/10.2139/ssrn.4976072.
10. Zhang, Q., Fang, C., Yu, B., Sun, W., Zhang, T., Chen, Z.: Pre-Trained Model-Based Automated Software Vulnerability Repair: How Far are We? IEEE Trans. Dependable and Secure Comput. 21, 2507–2525 (2024). https://doi.org/10.1109/TDSC.2023.3308897.
11. Roshanaei, M., Khan, M.R., Sylvester, N.N.: Enhancing Cybersecurity through AI and ML: Strategies, Challenges, and Future Directions. JIS. 15, 320–339 (2024). https://doi.org/10.4236/jis.2024.153019.
12. Kelvin Ovabor, Ismail Oluwatobiloba Sule-Odu, Travis Atkison, Adetutu Temitope Fabusoro, Joseph Oluwaseun Benedict: AI-driven threat intelligence for real-time cybersecurity: Frameworks, tools, and future directions. Open Access Res. J. Sci. Technol. 12, 040–048 (2024). https://doi.org/10.53022/oarjst.2024.12.2.0135.
13. Katiyar, Dr.N., Tripathi, Mr.S., Kumar, Mr.P., Verma, Mr.S., Sahu, Dr.A.K., Saxena, Dr.S.: AI and Cyber-Security: Enhancing threat detection and response with machine learning. eatp. (2024). https://doi.org/10.53555/kuey.v30i4.2377.
14. Rishit Lakhani: The Role of Artificial Intelligence in Enhancing Network Security: Opportunities and Challenges. Asian Journal of Multidisciplinary Research & Review. 4, (2023).
15. Kasri, W., Himeur, Y., Alkhazaleh, H.A., Tarapiah, S., Atalla, S., Mansoor, W., Al-Ahmad, H.: From Vulnerability to Defense: The Role of Large Language Models in Enhancing Cybersecurity. Computation. 13, 30 (2025). https://doi.org/10.3390/computation13020030.
16. Ajish, D.: The significance of artificial intelligence in zero trust technologies: a comprehensive review. Journal of Electrical Systems and Inf Technol. 11, 30 (2024). https://doi.org/10.1186/s43067-024-00155-z.
17. Gopireddy, S.R.: AI-Powered Code Review and Vulnerability Detection in DevOps Pipelines. (2024).
18. Samarakoon, S., Siriwardhana, Y., Porambage, P., Liyanage, M., Chang, S.-Y., Kim, J., Kim, J., Ylianttila, M.: 5G-NIDD: A Comprehensive Network Intrusion Detection Dataset Generated over 5G Wireless Network, https://arxiv.org/abs/2212.01298, (2022). https://doi.org/10.48550/ARXIV.2212.01298.
19. Ok, E., Eniola, J.: Best Practices for Cybersecurity Risk Management in Agile Software Development.
20. Vadde, B.C., Munagandla, V.B.: AI-Driven Automation in DevOps: Enhancing Continuous Integration and Deployment. 01, (2022).
21. Jimmy, F.: Emerging Threats: The Latest Cybersecurity Risks and the Role of Artificial Intelligence in Enhancing Cybersecurity Defenses. int.jour.sci.res.mana. 9, 564–574 (2021). https://doi.org/10.18535/ijsrm/v9i2.ec01.
22. Reddy Julakanti, S.: AI Techniques to Counter Information Security Attacks. IJRITCC. 11, 518–527 (2023). https://doi.org/10.17762/ijritcc.v11i5.11368.
23. Kyler, T.: AI-Driven DevSecOps: Integrating Security into Continuous Integration and Deployment Pipelines.
24. Vinothkumar, J., Karunamurthy, A.: Recent Advancements in Artificial Intelligence Technology: Trends and Implications. qijmsrd. 2, 1–11 (2023). https://doi.org/10.54368/qijmsrd.2.1.0003.

A Comparative Analysis of Dynamic and Static Malware Detection Tools

Haseeb Ahmed, Hasnat Ali, Komal Batool, Muhammad Islam, and Azeem Irshad

Abstract Malware poses a significant threat to the modern computer world, and various tools and techniques exist for detecting malware. This research article aims to address this challenge by presenting a comprehensive performance comparison of existing malware detection tools and techniques. Moreover, in this paper, a performance comparison of existing tools and techniques for malware detection is presented. To assess the performance of these tools, a virtual malware analysis lab is created using VMware workstation. We have selected 17 widely used malware detection tools and utilized known malware samples for our comparison. Through rigorous testing and analysis, we evaluate the performance of these tools based on multiple parameters, i.e., Accuracy, Specificity, Sensitivity, and Efficiency which are also presented graphically. The findings indicate that the top three tools, based on specific parameters and the given dataset, are Process monitor, IDA freeware, and Intezer. These tools demonstrate remarkable performance in terms of detection accuracy, low false positive rates, efficient detection speed, optimized resource utilization, and intuitive user interfaces. This performance comparison study contributes to the field of malware detection by providing valuable insights into the strengths

H. Ahmed
Department of Computer Science and Cyber Security, Air University Islamabad, Islamabad, Pakistan

H. Ali · K. Batool
Faculty of Computing and Informatics, Multimedia University, Cyberjaya, Malaysia

Faculty of Computing and Informatics, Riphah International University, Islamabad, Pakistan

H. Ali
e-mail: hasnat.ali@riphah.edu.pk

K. Batool
e-mail: komal.batool@riphah.edu.pk

M. Islam
College of Science and Engineering, James Cook University, Cairns, Australia
e-mail: muhammad.islam1@my.jcu.edu.au

A. Irshad (✉)
PHED, GGC Asghar Mall Rawalpindi, Rawalpindi, Pakistan
e-mail: azeemirshad@gpgcam.edu.pk

M. Azrour and A. Alabdulatif (eds.), *AI-Driven Security for Next-Generation IoT Systems*, https://doi.org/10.1007/978-3-032-08784-3_7

and limitations of different tools and techniques. The results can aid security professionals, researchers, and developers in making informed decisions when selecting and implementing malware detection solutions.

Keywords Malware · Indicator of compromise · Command and control server · Dynamic link library · Zero-day behavior-based · Signature-based · Online malware detection

1 Introduction

As we are living in a type of era where everything has been digitalized. Along with this emergence, we are also facing many problems in our daily life in the form of malware. It is very important to make sure that our computer programs are malware-free and there is no damage caused by the malware. Malware is increasing day by day and in every phase, we should make something concerned that we should be protected from malware. All software variants like viruses, Trojans, ransomware, spyware, adware and Trojans are considered as malware. The focus of creating such malicious malware is to gain unauthorized access to a network. More damages of such malware include deleting files, the stealth of data, and corruption of files. From the stealth of data to gaining personal information, all of this is considered as bad intent of specific malware. Cybercriminals are using advanced methods in this aspect and are using more sophisticated techniques of infecting a system. To protect users and corporate space it is mandatory to identify such malware. Malware detection is the process to identify if a specific program has some malicious intentions or not. The advancement of different obscured zero-day malware consistently inspired security analysts to adapt strategies to alleviate using malware detection tools. The objective of detection tools is to identify what happens and to make sure that you find all infected programs and files. In the old days, signature-based techniques were used to identify malware, but this technique has a certain limitation: it cannot detect zero-day malware. Zero-day malware can be identified using machine learning techniques. There is also another method of identifying such malware called the Heuristic-based technique, but it also has the limitation of giving false positive rates. It is not possible to eliminate all malware as it is increasing day by day and it is a tedious task to delete all malware. The entry points of malware can be any system including Windows, Linux, Android, and Cloud systems.

Malware authors try to avoid detection using polymorphic and metaphoric techniques and methods like Dead-Code Insertion, Register Reassignment, Subroutine Reordering, Instruction Subroutines Code Transportation, Code Integration, and Concealing Strategies. All those software and computer programs are considered malware that causes harm to users. The fundamental objective of this research article is to inspect and decide about probably the most well-known malware recognition tools by comparing their results. We have created a virtual environment to deal with these problems to protect the rest of our system from malicious activities. After

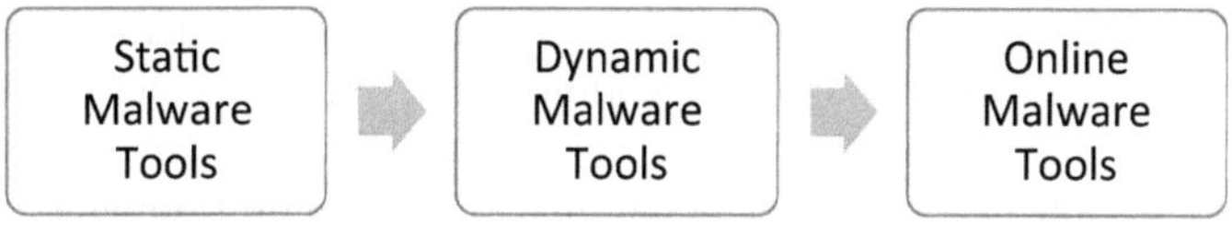

Fig. 1 Approaches used in this research

this, analyzing datasets on malware analysis tools a comparison report of different malware tools is based on the results which have given a brief idea about the performance of malware analysis tools. The malware tools used in this paper are shown in Fig. 1.

2 Malware Analysis Techniques

This section will provide detailed information about the current malware identification and investigation techniques. Mainly malware analysis techniques are divided into two broad categories static and dynamic analysis while hybrid analysis is the fusion of static and dynamic malware analysis techniques.

1. Static malware analysis [1]
2. Dynamic malware analysis [2]
3. Hybrid malware analysis [3]

2.1 Static Malware Analysis

It is a method to judge the behavior of a program without execution. Using this technique no harmful behavior will affect the specific device. It is often performed to check the vulnerabilities of the system and to deal with performance issues. Static code analysis is a technique of troubleshooting by inspecting source code before a program is run [4]. This process is done by examining a set of code against a set (or various sets) of coding rules. All static tools search specific code to find patterns that are harmful and have some sort of warning information with them. There are specific tools [5] available for static analysis some of which we will apply in this research for code inspection and sensitive feature identification.

2.2 Dynamic Malware Analysis

Dynamic analysis is the act of progressively examining software while it is running. Basic dynamic analysis is the method involved with testing and assessing a program while the product is running. Likewise, this analysis is code examining, dynamic investigation works on the analysis and rectification of bugs, memory issues, and crashes of an application during its execution [6]. Several tools and sandboxes are available for dynamic analysis in a virtual environment. Whereas progressed dynamic analysis utilizes a debugger to review the inner condition of a running malevolent application.

2.3 *Hybrid Malware Analysis*

To cover the limitations of both methodologies hybrid analysis is used. Certain activities that can be concealed at run-time might be recognized while unpacking the binary files or reviewing them in assembly code [7]. To understand advanced parts of analysis this technique is used and it provides a better understanding of hidden malware.

3 Related Work

On 03 May 2023, ransomware gang hijacks university alert system to issue threats. All exams had to be postponed due to the impact on the IT systems and inform the students that they had exfiltrated 1.2 TB of files, including thousands of students' admission and personal information, from the university network [8]. According to Seqrite Threat Report 2021, Over 1.9 million ransomware assaults, up 15% from the previous year, according to the research [9]. On 02 July 2023, a malware attack on the American software company Kaseya caused the services of thousands of organizations to be interrupted [11]. In a survey paper [12], a complete overview of malware detection tools is presented but it has limitations as no such boundary is defined on basis of which anyone can make a decision on which malware tool will be considered as best for malware detection. Similarly, in a paper [13], detailed knowledge of advanced malware and detection methods are discussed. But there is no practical implementation based on which someone decides about the performance of malware tools. Machine learning is considered the best approach for detection mechanism in a recent study [14] malware detection through ML, the technique is considered the best approach to deal with sophisticated malware but it has also certain limitations as it is not possible to make a decision which tools are better for malware analysis. For android malware detection a survey is conducted [15] to tell which techniques are available for malware detection mechanisms. But no such scenario is discussed which tells about the efficiency of malware analysis tools.

The concept of hybrid analysis is sometimes considered the best approach for detection mechanisms. Hybrid analysis means combining both static and dynamic methodology for malware detection. As in paper [16] for android malware detection and protection, hybrid analysis is considered a better approach for detection. A limitation of this research is the lack of practical implementation, and no rules defined to decide on the best available malware tools for analysis. A malware family is a program or set of associated programs with enough code overlap to be considered part of the same group [17]. In a survey of windows/desktop malware, it is concluded that malware detection accuracy can be improved by separating a group of malware families with their functionality [18]. This technique is considered the best to detect second-generation malware. The limitation of this research is no information about malware tools is provided that which tool performs best to deal with such situations.

A framework for dynamic analysis using cuckoo is presented in this paper [19]. The limitation of this work is focusing on the dynamic analysis approach using the cuckoo sandbox. More dynamic analysis tools should be added to check their results and compare the results of tools according to them. A structure for comparative examination of benign and malignant android applications, given static highlights, can be utilized [20]. The analysis is based on permission-based with specific malware family classification. This feature has a limitation of static tools comparison and the addition of dynamic analysis features. A study of the classifying approach by studying the API calls together [21] with their arguments is used. The methodology has guaranteed wide endorsement in API-based malignant behavior recognition for the Windows platform, but this technique has a limitation as a comparison of malware tools is missing.

A comparison of malware tools is applied [22] which is a good approach to finding the most suitable static and dynamic analysis tools. A report is generated and an Intezer analyzer is considered the best malware tool for analysis. But this study has limitations. The data set that is used is very limited. All experiments are performed on only one data set. For more accurate results different data sets should be tested to make a better understanding of the performance of malware tools. In a paper [23] large data sets are collected, and the PE header of files is analyzed. It is concluded that the PE header is much faster than scanning the whole information in the PE files. But it has a limitation as only analyzing the PE header does not give all details of specific malware. Moreover, a framework that compares efficient PE analyzing tools is missing and it is not mentioned which tool can give a better result.

A comparison of malware tools is presented in this research work [24]. Different data sets are used, and static and dynamic techniques are applied. For this purpose, a virtual environment created for malware analysis. Finally, the top three tools are decided best tools for malware analysis based on specific parameters.

This paper is focused on analyzing malware tools and presents a comparison report which will give a clear idea about the performance of specific malware tools.

4 Malware Detection Techniques

In this section, we will discuss malware detection techniques, i.e., signature-based, behavior-based, and heuristic-based detection mechanisms.

4.1 Signature-Based Detection

Identifying the known malware with the help of its footprints is called a signature-based detection method. All types of software and programs have unique digital footprints or signatures present in their code. We can easily detect a specific malware using its signature. If we talk about antivirus products all of them use this technique for decades to find footprints of known malware. All types of footprints are stored in the database and if some malware characteristic matches with the record they declare it as malware. Whenever new malware is found by security vendors they add that specific footprint to the database. However, it has certain limitations as it cannot detect zero-day malware attacks. Signature-based cannot discover new or previously hidden malware and just modifying the signature of a malware file can completely bypass the detection mechanism. New-generation advanced malware can change its signatures to avoid detection phenomena. If the signature is missing in the respective database record the attack will be unable to detect by detection tools.

4.2 Behavior-Based Detection

In this technique, the behavior of a program tells the real purpose of that program file. A specific object/file is executed in an environment to check its behavior. The unauthorized or abnormal behavior decides whether the program file is malicious or not. This technique has also some limitations as malware is executed in the sandbox or virtual environment, it is very important that malware cannot be detected in the virtual environment because it is possible that malware does not show full intent and hides its behavior. This can lead to wrong malware analysis and required results cannot be found.

5 Malware Detection Tools

In this section, we will describe the features and limitations of malware analysis tools in detail.

We divided malware detection into three categories which are the following.

1. Static malware detection tools
2. Dynamic malware detection tools
3. Online malware detection tools

Table 1 Static malware analysis tools

S. No.	Tools	Detection method
01	PeStudio	Signature-based approach
03	PEID	Signature-based approach
03	Dependency walker	Dependency analysis
04	Resource Hacker	Signature-based approach
05	CFF Explorer	Signature-based approach
06	ExeInfo PE	Signature-based approach
07	IDA Freeware	String matching

Table 2 Dynamic malware analysis tools

S. No.	Tools	Detection method
01	Process explorer	Behavior-based approach
02	Process hacker	Behavior-based approach
03	OllyDbg	Heuristic-based approach
04	Regshot	String matching
05	Process monitor	Behavior-based approach
06	Cuckoo sandbox	Behavior-based approach

5.1 Static Malware Detection Tools

In static malware analysis, we analyze the malware without execution which is helpful in identifying potential security threats in a sample without the risk of infecting the analysis environment. List of static malware detection tools that we used for analyzing malware can be shown in Table 1.

5.2 Dynamic Malware Detection Tools

In dynamic malware analysis, we analyze the behavior of malware after executing it in a safe environment which helps malware analysts to understand the true nature of a threat.

The list of dynamic tools for malware analysis is shown in Table 2.

5.3 Online Malware Detection Tools

It is fully automated malware analysis which gives information about suspicious URL's and files. Table 3 shows some tools listed for online malware analysis.

Table 3 Online malware analysis tools

S. No.	Tools	Detection method
01	Virus Total	Heuristic-based approach
02	Any. Run	Signature-based approach
03	Cuckoo cert. ee	Behavior-based approach
04	Intezer	Genetic-based approach

6 Steps for Static Malware Analysis

Steps involved in static malware analysis are:

Figure 2 shows all steps involved in static malware analysis and these steps will extract valuable information related to malware samples.

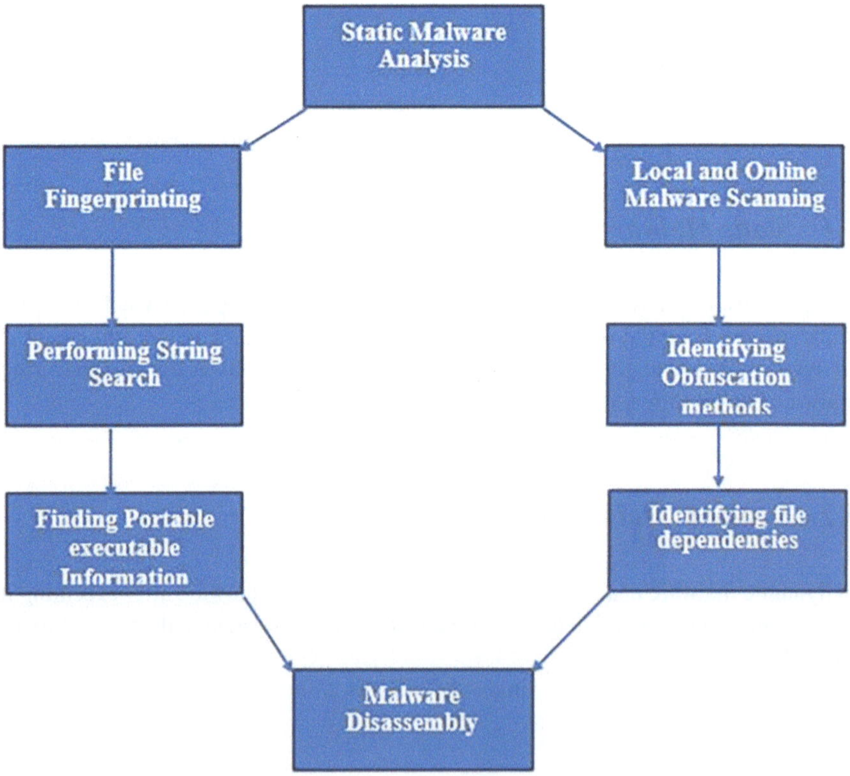

Fig. 2 Static malware analysis steps

7 Proposed Methodology

To know which tool will perform best we have created a virtual environment, and we will analyze malware on various data sets based on that analysis we will compare the execution of different malware tools.

- We have taken malware samples from Dike data set [25] and used public submission reports and sample from Any. Run [26] for our comparison report.
- We tested and analyzed the malware samples using various tools to check the performance and limitations of each tool.
- It has given comprehensive knowledge about the performance of malware tools and techniques that researchers can use for further studies.

The first step of malware analysis is to ta take a snapshot of the system to keep a backup of the system and it is a way of protecting your data and allowing the application to write their data. Next, we will decide on which environment we want to execute the malware sample, i.e., Virtual Environment. After deciding on the specific environment first, we will analyze malware statically and then we allow malware to execute and check the behavior of malware dynamically.

After execution, we stop the analysis tool, and the final step of analysis is to analyze the sample and check the result of the whole process. After analyzing, we will again take a new malware sample and restore the system. After restoring the same process is repeated and after desired analysis, we will end the process and compare the performance of malware tools. Figure 3 shows a flow diagram of malware analysis that describes the complete process of the malware analysis technique. Below, we can see a list of static, dynamic, and online malware analysis tools. We will analyze our data sets using some of these tools and compare their results to check the performance of our respective tools. The list of Static malware analysis tools for malware analysis is shown in Table 1.

8 Experimental Results

Experimenting on malware samples can be a complex and potentially dangerous task, so it's important to approach it with caution and appropriate expertise. First and foremost, it's essential to set up a secure environment to run the malware samples. This could involve creating a virtual machine or container that's isolated from the rest of your system and network, using sandboxing tools or specialized analysis software. Before running the malware, you'll want to gather information about its behavior, characteristics, and potential impact. This could involve analyzing the code, looking for known signatures or indicators of compromise, and researching any existing reports or analyses of the malware. For this purpose, we will analyze malware samples using static and online malware tools. Once you've set up a secure environment and done an initial analysis, you can run the malware sample to observe

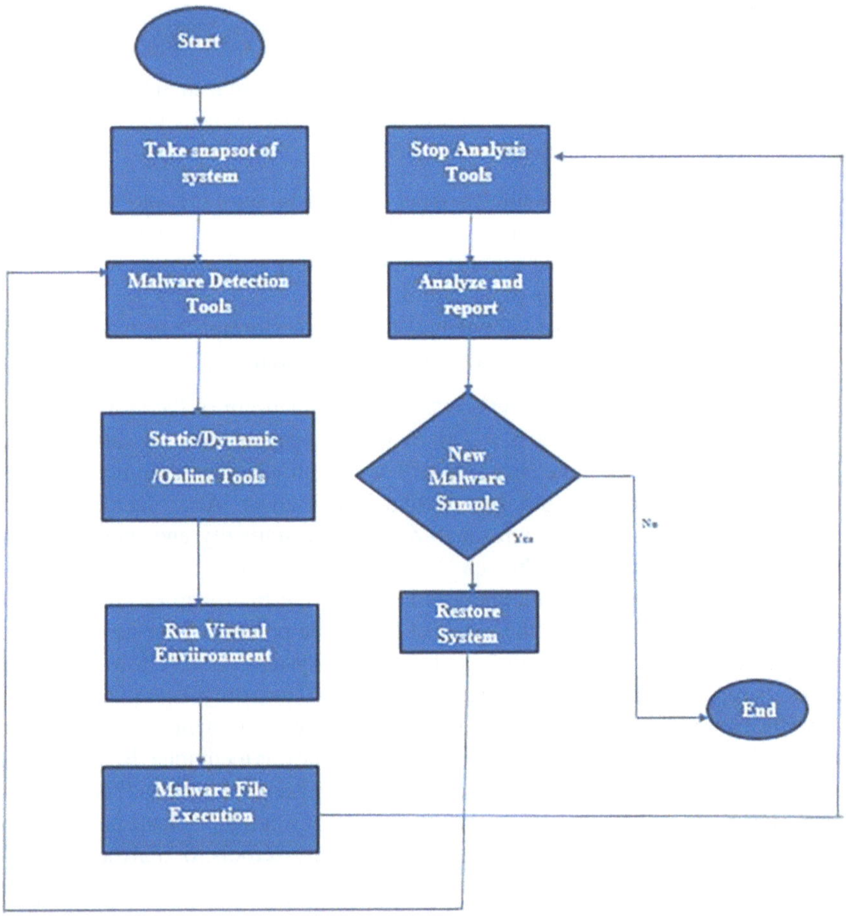

Fig. 3 Malware analysis workflow sequence

its behavior. You may want to monitor system calls, network traffic, and other indicators of activity to see what the malware is doing. Here, we will take advantage of using dynamic malware analysis tools to analyze behavior of specific malware samples. After running the malware, analyze the results of the experiment. To gain a better understanding of the malware's behavior and potential impact, repeat the experiment with variations or different configurations. This can help to refine analysis and identify new indicators of compromise or potential vulnerabilities. Remember, experimenting on malware samples can be risky, so it's important to take appropriate precautions and seek expert guidance if needed. Additionally, it's important to use only legally obtained malware samples and to follow all applicable laws and regulations. While comparing performance of malware detection tools, the 4 main factors listed below are considered.

8.1 Sensitivity

Sensitivity refers to the capacity of a tool to accurately identify a given condition. Tools with higher sensitivity values are considered to have better performance. Sensitivity can be defined as the proportion of true positive results (i.e., correctly identified cases) out of all actual positive cases.

$$\text{Sensitivity} = \frac{\text{No of TP}}{\text{No of TP+No of FN}} \tag{1}$$

In general, the calculation of sensitivity, specificity, accuracy, and efficiency considers the following four terms: true positives (TP), false positives (FP), true negatives (TN), and false negatives (FN). These terms are used to determine the tool's ability to correctly identify positive cases (TP) while minimizing the identification of false positives (FP) and false negatives (FN) and maximizing the identification of true negatives (TN).

8.2 Specificity

Specificity refers to the ability of a tool to correctly exclude individuals who do not have the condition being tested. A lower value of specificity indicates better performance of the tool. Specificity can be expressed as the proportion of true negative results (i.e., correctly identified cases where the condition is absent) out of all actual negative cases.

$$\text{Specificity} = \frac{\text{No of TN}}{\text{No of TN+No of FP}} \tag{2}$$

8.3 Accuracy

Accuracy refers to the proportion of correct results (both true positives and true negatives) in a given sample. Higher values of accuracy indicate better performance of the tools.

Where True Positives (TP): the number of malware samples correctly identified as malware. False Positives (FP): the number of benign samples incorrectly identified as malware. True Negatives (TN): the number of benign samples correctly identified as benign. False Negatives (FN): the number of malware samples incorrectly identified as benign. The result will be a percentage value that indicates the overall accuracy of the malware tool in correctly identifying malicious and benign samples.

$$\text{Accuracy} = \frac{\text{Noof(TP+TN)}}{\text{Noof(TP+FP+TN+FN)}} \tag{3}$$

8.4 Efficiency

The efficiency of malware tools refers to their ability to effectively detect, analyze, and mitigate malware threats while utilizing minimal system resources and minimizing false positives or false negatives. The tools will perform better if their efficiency value will be higher (Table 4).

To evaluate the performance of malware tools four important parameters are considered which were Accuracy, Sensitivity, Specificity, and Efficiency. Based on experimental results and the given dataset, top three tools to detect malware are IDA Freeware for static analysis, Process explorer for dynamic analysis and Intezer for online malware analysis. Table 5 is showing complete details of performance measures of different static, dynamic and online malware tools.

After analyzing malware samples using various tools, we made a comparison table which shows the performance of each tool using different parameters. We categorize malware tools in three different categories to make performance comparison separately of each detection method, i.e., static malware tools, dynamic malware tools, and online malware tools. Several things are included to measure efficiency

Table 4 Malware samples

S. No.	Malware file name	Malware family
01	all in one checker.exe	Adware
02	Keylemon_rockbytes.exe	Adware
03	RIP_YOUR_PC_LOL.exe	Backdoor
04	ProviderReviewNet.exe	Backdoor
05	LockBit_14_02_2021_146KB.exe	Ransomware
06	wannaCry.exe	Ransomware
07	Orderlist.exe	Trojan
08	Mw2Loader.exe	Rat
09	OnlineInstaller.exe	Rootkit
10	Setup.exe	Downloader
11	Sad.exe	Stealer
12	all in one checker.exe	Adware
13	ProviderReviewNet.exe	Backdoor
14	W4TCH3D.exe	Ransomware
15	wannaCry.exe	Ransomware
16	file.exe	Trojan
17	Mw2Loader.exe	Rat
18	OnlineInstaller.exe	Rootkit
19	Setup.exe	Downloader
20	Sad.exe	Stealer
21	Azure Fresh Card.exe	Trojan

Table 5 Performance comparison of malware detection tools

S. No.	Name of tools	Total malware samples	Detected as malware	Detected as benign	Accuracy (%)	Sensitivity (%)	Specificity (%)	Efficiency (%)
01	VirusTotal	31	16	15	80.64	91.66	42.85	51.61
02	Any. Run	31	17	16	83.87	91.66	57.14	54.83
03	Cuckoocert.ee	31	18	13	77.41	91.30	37.5	54.83
04	Intezer	31	17	14	83.87	92.30	34.86	58.06
05	Process explorer	31	24	7	84.84	96.29	33.33	74.19
06	Process monitor	31	23	8	87.09	96.29	25	77.41
07	Regshot	31	22	9	83.87	95.83	42.85	74.19
08	Cuckoo sandbox	31	21	11	80.64	88.46	40	67.74
09	OllyDbg	31	11	19	70.96	89.47	41.66	35.48
10	Process Hacker	31	21	10	77.41	95.00	45.45	67.7
11	PeStudio	31	18	14	70.96	77.27	55.55	58.065
12	PEID	31	17	11	77.41	91.66	28.57	54.83
13	IDA Freeware	31	19	12	79.31	91.30	33.33	61.29
14	Resource Hacker	31	18	13	71.875	82.60	44.44%	58.06%
15	CFF Explorer	31	12	19	70.96	77.27	55.55	38.70
16	ExeInfo PE	31	20	11	73.33	86.36	37.51	64.51
17	Dependency Walker	31	15	16	74.19	79.16	57.14	48.38

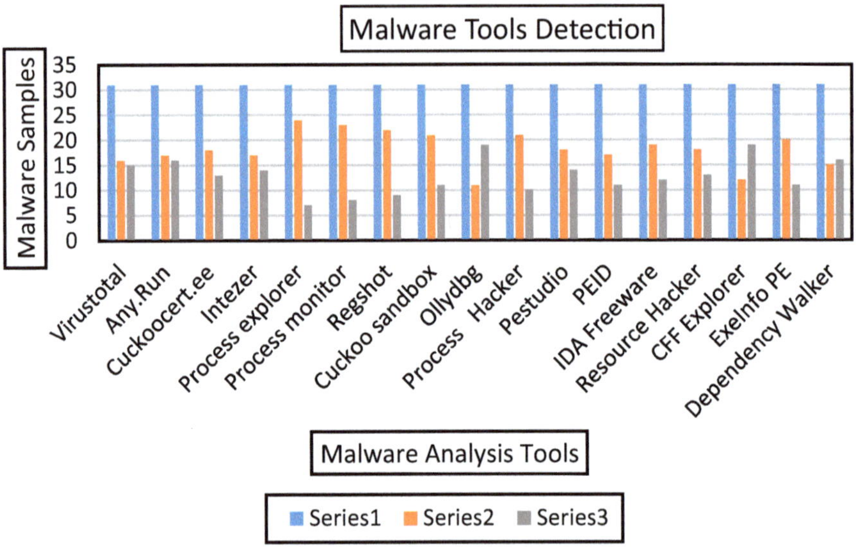

Fig. 4 Malware tools detection

of malware tool, i.e., detection rate, low false positive ratio, low false negative ratio, resource usage and analysis time. Using all these parameters a performance comparison table of the best available tool for malware analysis is decided as shown in Table 5.

In Fig. 4, we can see the total number of malware samples which are detected as malicious and benign by malware tools. Malware detection tools identify malware samples as malicious or benign using a variety of approaches and techniques. These techniques can include signature-based detection, behavior analysis, sandboxing, heuristic-based string matching, etc.

In Fig. 5 we can see malware tools analysis ratios obtained from the analysis of malware samples using various malware analysis tools or techniques. Analysis ratios can include various metrices, i.e., true positive, true negative, false positive, false negative ratios. These ratios can offer valuable information on the functionality and effectiveness of malware analysis tools, enabling researchers, security experts, and organizations in making informed decisions about strategies used in tool set.

In Fig. 5 we can see the accuracy factor of malware analysis tools on various malware samples. It serves as an indicator of the tool of overall accuracy in detecting and finding malware. A malware tool that has a higher precision factor is more reliable and precise. The ongoing evolution and sophistication of malware make it incredibly difficult to get a 100% accuracy rate, it is important to highlight. The effectiveness of the detection algorithms, the size and quality of the malware signature database, and the amount of tool updates are other variables that can affect accuracy.

In Fig. 6 sensitivity of malware tools is identified, and it refers to its ability to precisely detect and categorize malicious files or activities without generating an

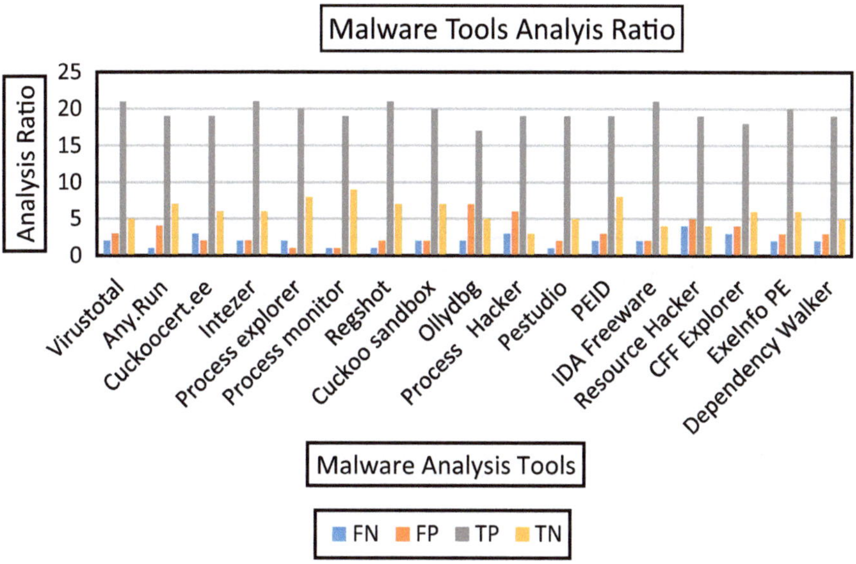

Fig. 5 Malware tools analysis ratio

excessive number of false negatives. Even if malware is new or has been altered to prevent detection, a malware tool with high sensitivity is more likely to find it. However, a malware tool with high sensitivity may also be more likely to generate false positives, which are alerts that are triggered by benign software. A sensitive malware tool has a low rate of false detections and can recognize a variety of malware samples, including both known and unknown varieties.

In Fig. 7 specificity of malware tools can be seen, and it refers to its ability to precisely detect and categorize malicious files or activities without generating an excessive number of false positives. Only harmful files will be reported by a particular malware tool, which can efficiently distinguish between legitimate software and potential threats. False positives may result in unnecessary inconvenience, lost productivity, or other negative outcomes.

The ability of a malware tool to carry out its specified tasks promptly and with minimal resource consumption is referred to as efficiency. An effective malware tool enhances performance and effectiveness while utilizing the least amount of CPU, memory, and storage space possible. We can see the efficiency of malware tools identified in Fig. 8.

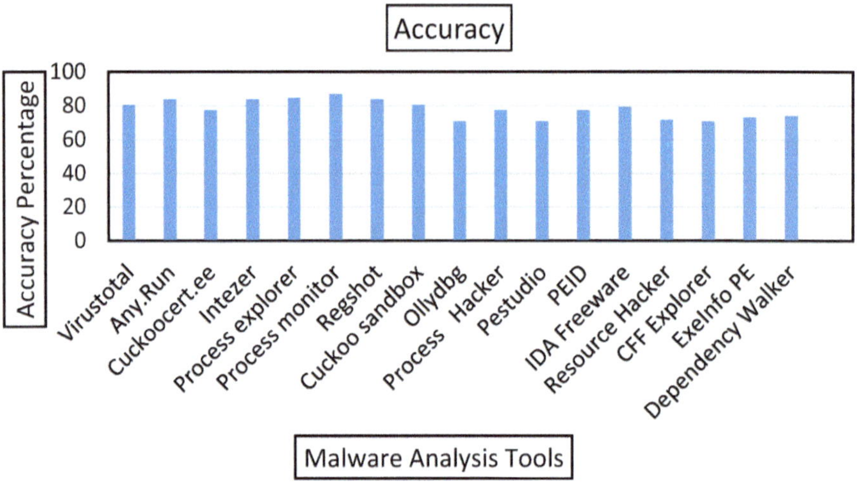

Fig. 6 Accuracy of malware tools

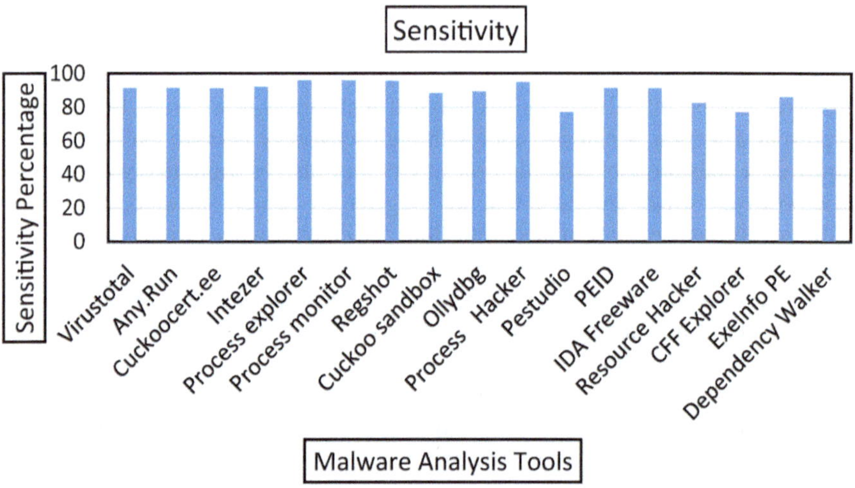

Fig. 7 Sensitivity of malware tools

9 Comparison Table of Strength and Weaknesses of Malware Tools

In this section the strengths and weaknesses of malware tools are discussed in detail (Fig. 9).

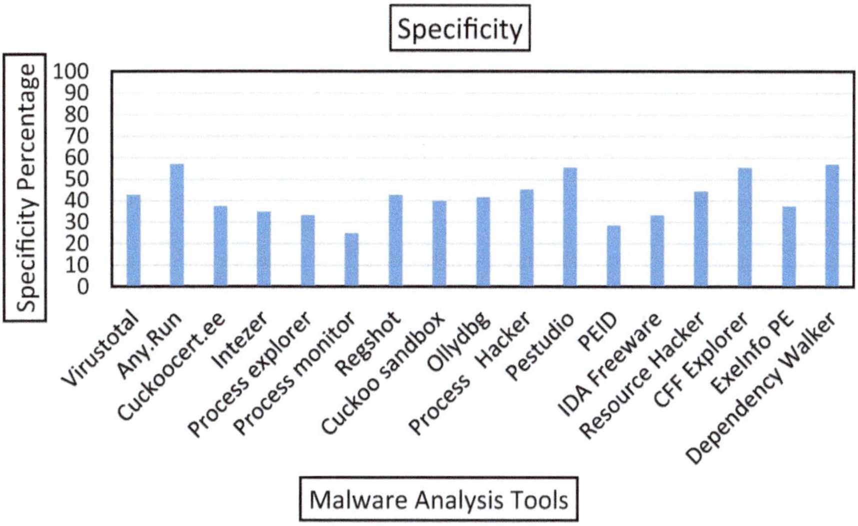

Fig. 8 Specificity of malware tools

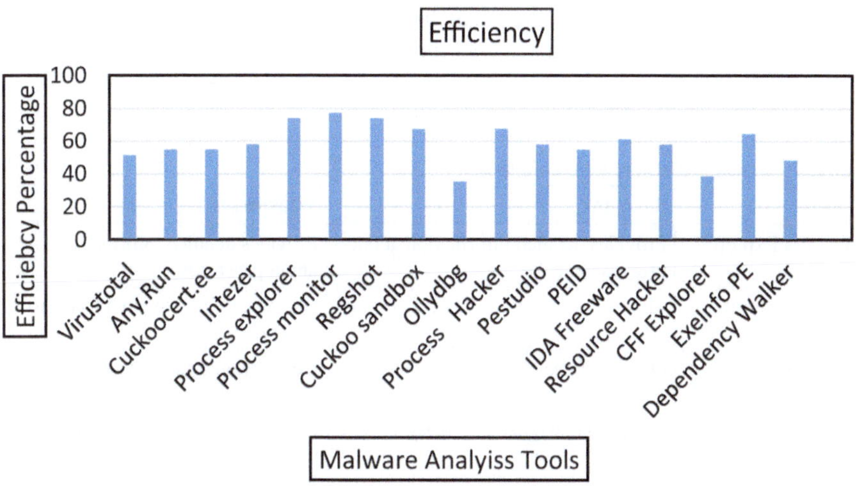

Fig. 9 Efficiency of malware tools

9.1 Static Malware Tools Comparison

Table 6 is showing performance table outlining the strengths and weaknesses of the static analysis tools, i.e., PEID, PEStudio, Dependency Walker, Resource Hacker, CFF Explorer, Exeinfo PE, and IDA Freeware.

Table 6 Static malware tools comparison

Tools	Strengths	Weaknesses
PEID	Identify packers and protectors	Limited file format support
PEStudio	Provides detailed information and analysis of PE files	Lacks advanced disassembly and debugging capabilities
Dependency Walker	Displays dependencies of a PE file	Limited malware-specific analysis capabilities
Resource Hacker	Allows modification of resource data	Focused on resource editing, lacks comprehensive analysis
CFF Explorer	Provides detailed PE file inspection	May not detect advanced obfuscation techniques
Exeinfo Pe	Detects packers, compilers, and file types	May generate false positives
IDA Freeware	Advanced disassembly and code analysis	Steep learning curve, primarily for expert users

9.2 Dynamic Malware Tools Comparison

A performance table outlining the strengths and weaknesses of the dynamic analysis tools is shown in Table 7 and selected tools for comparison are Process Explorer, Process Monitor, Process Hacker, OllyDbg, Regshot, and Cuckoo Sandbox.

Table 7 Dynamic malware tools comparison

Tools	Strengths	Weaknesses
Process Explorer	Provides detailed information about processes	Provides detailed information about processes
Process Monitor	Monitors process and system activity	Overwhelming amount of data to analyze
Process Hacker	Offers advanced process and system analysis	Complex interface, may be challenging for beginners
OllyDbg	Powerful debugger for binary analysis	Steep learning curve, primarily for expert users
Regshot	Compares before-and-after registry snapshots	Focuses only on registry changes, limited malware analysis
Cuckoo Sandbox	Automated malware analysis and sandboxing	Requires setup and configuration, resource-intensive

Table 8 Static malware tools comparison

Tools	Strengths	Weaknesses
VirusTotal	Aggregates results from multiple scanners	Reliance on just signature-based analysis
Any. Run	Provides dynamic behavior analysis	Limited free usage requires an account for full functionality
Intezer	Analyzes and detects code reuse in malware	Limited visibility into dynamic behavior
Cuckoocert.ee	Offers malware analysis and reporting	Limited Availability and access

9.3 Online Malware Tools Comparison

Similarly, the performance table outlines the strengths and weaknesses of online analysis tools, i.e., VirusTotal, Any. Run, Intezer, and Cuckoocert.ee. can be shown in Table 8.

10 Conclusion and Future Work

Malware is one of the most common methods of cyberattacks. But there are steps you can take to make yourself less vulnerable. By doing what you can to educate yourself, you're less likely to become a victim of cybercrimes. This information can be utilized to create new and improved malware detection systems. Researchers can develop new methods of malware detection that are more effective than either static or dynamic analysis alone by combining the strengths of static and dynamic analysis tools. By comparing the results of static and dynamic analysis tools, researchers can improve the accuracy of malware detection. This is because each approach can detect different types of malwares. By comparing the results of static and dynamic analysis tools, researchers can develop new tools for malware analysis. These tools can be used to automate the process of malware analysis and to make it easier for analysts to identify and respond to malware threats. To investigate existing malware detection methods and methodologies, a virtual environment was created for malware analysis. Malware samples are thoroughly tested and analyzed using effective malware detection techniques. Based on experimental results and the given dataset, it is decided that top three tools to detect malware in terms of performance based are IDA Freeware, Process explorer and Intezer. However, we can never say about a specific tool that it is ideally best tool for all types of malware analysis. It is always recommended to use a combination of static and dynamic malware tools to enhance our results.

A potential avenue for future research in this field is the development of a new technique by integrating various analysis tools into a unified framework. This technique can improve the strengths and capabilities of each tool, effectively combining their unique metrics and analysis methodologies. This holistic approach would enable a more comprehensive analysis of malicious code and improve the accuracy and reliability of detection. Additionally, the technique could be designed to continuously evolve and adapt to emerging threats by integrating real-time threat intelligence feeds and leveraging machine learning techniques. This would enhance its capability to detect new and evolving malware variants more accurately and rapidly. Overall, the development of a technique that is the collective power of multiple malware analysis tools has the potential to significantly advance the field of malware analysis, improving system protection and strengthening cybersecurity defenses.

References

1. A. Richardson, "Static Analysis," *SecureCodeWarrior,* 1 February 2022 Available at: https://www.securecodewarrior.com/blog/ [Online; accessed 10-January-2022].
2. Total View, "Dynamic Analysis," 10 October 2020 Available at: https://totalview.io/blog/what-dynamic-analysis [Online; accessed 10-January-2022].
3. Lifars, *Difference between Static, Dynamic and Hybrid Malware Analysis,* 11 April 2020 Available at: https://lifars.com/ [Online; accessed 13-January-2022].
4. R. Bellairs, "Static Analysis Security And Compliance," What is Static Analysis Code An Overview, Feb. 10, 2020. [Online]. Available: https://www.perforce.com/blog/sca/what-static-analysis. [Accessed: Jan. 17, 2020].
5. 0*1 Agency, "0*1 Cyber Of Security," *Malware Analysis Tools List,* 2022 Available at: https://0x1.gitlab.io/security/Malware-Analysis-Tools-List/ [Online; accessed 23-January-2022].
6. Total View, "Dynamic Code Analysis," 10 July 2020 Available at: https://totalview.io/blog/what-dynamic- [Online; accessed 25-January-2022].
7. Lifars, "What is hybrid Analysis," *Difference between Malware analysis Techniques, [Online; accessed* 11 November 2020] Available at: https://lifars.com/2020/11/what-are-the-differences-between-static-dynamic-and-hybrid-malware- [Online; accessed 28-January-2022].
8. CM Alliance. (2023, May). Recent Cyber Attacks, Data Breaches & Ransomware Attacks. CM Alliance Cybersecurity Blog. Retrieved from https://www.cm-alliance.com/cybersecurity-blog/may-2023-recent-cyber-attacks-data-breaches-ransomware-attacks
9. Seqrite. (2022). Threat Reports. Retrieved from https://www.seqrite.com/resources/cat/threat-reports [Accessed: July. 18, 2023]
10. IDC Corporate, "Smartphone Market Share," *Supply Chain Constraints Smart Phone Market,* 28 October 2021Availableat: https://www.idc.com/promo/smartphone-market-share/os [Online; accessed 4-February-2022].
11. Wikipedia. (2023, July 18). Kaseya VSA ransomware attack. In "Wikipedia". Retrieved July 18, 2023,from https://en.wikipedia.org/wiki/Kaseya_ransomware
12. S. Talukder and Z. Talukder, "A Survey on Malware Detection and Analysis Tools," *International Journal of Network Security & Its Applications,* vol. 12, 2020.
13. B. M. Mehtre and Heena, "Advances In Malware Detection-An Overview," vol. II, 2021.
14. O. Abikoye, B. Gyunka, and A. OLUWATOBI, "Android Malware Detection through Machine Learning Techniques: A Review," *International Journal of Online and Biomedical Engineering (iJOE),* vol. 16, p. 14, 2020.
15. M. Odusami, O. Abayomi-All, S. Misra and O. Shobayo, "Android Malware Detection: A Survey," *International Conference on Applied Informatics, ICAI 2018At: Bogota, Colombia,* 2018.
16. S. Arshad, M. A. Shah, A. Khan, and M. Ahmed, "Android Malware Detection & Protection: A Survey," *International Journal of Advanced Computer Science and Applications,* vol. 7, 2016.

17. Fireeye, "Malware Family Classification". *FireEye Mandiant Researches Malware Families Available at:* https://vision.fireeye.com/.
18. S. Sahay and A. Sharma, "A Survey on the Detection of Windows Desktops Malware," 2019. [Online; accessed 02-December-2021].
19. K. Navroop, A. K. Bindal, and A. PHD, "A complete dynamic malware analysis.," *International Journal of Computer Applications 135,* vol. IV, pp. 20–25, 2016.
20. N. Chavan, F. D. Troia, and M. Stamp, "A Comparative Analysis of Android Malware," 2019.
21. D. Rabadi and S. Teo, "Advanced Windows Methods on Malware Detection and Classification," *Annual Computer Security Applications Conference,* pp. 54–68, 2020. [Online; accessed 10-February-2022].
22. A. Datta, K. A. Kakelli and A. D, "An Emerging Malware Analysis Techniques and Tools: A Comparative Analysis," 2021.
23. S. Kim, "PE header analysis for malware detection," 2018. *Computer Science and Applications,* Vols. 11, No 7, 2020.
24. K. S. Pandey and M. B. M., "Performance of malware detection tools: A Comparison," in *2014 IEEE International Conference on Advanced Communications, Control and Computing Technologies*, 2014. [Online; accessed 15-February-2022].
25. I. Ache, "DikeDataset," GitHub. [Online]. Available: https://github.com/iosifache/DikeDataset. [Accessed: May 30, 2023].
26. M. Naveed, "Intrusion Detection in Smart IoT Devices for People with Disabilities," 2022 IEEE International Smart Cities Conference (ISC2), Pafos, Cyprus, 2022.

Network Behavior–Driven Intrusion Detection: A Hybrid Deep Learning and XGBoost Approach

Mohammed Janati and Fayçal Messaoudi

Abstract New challenges in the cyberspace require the need for more powerful and rapidly adaptive Intrusion Detection Systems (IDS). This paper presents a hybrid deep learning method—merging autoencoders with XGBoost, extending with Network Behavior Analysis (NBA) feature extraction and chi-squared–based feature selection and evaluated on TII-SSRC-23 dataset. The raw network traffic will be preprocessed and enriched with innovations inspired from NBA features, like aggregate session- and flow-level statistics (e.g., top numerical features sum, mean of odd-indexed values). Then, we execute a chi-squared test to weed out the least relevant; dimensionality is now trimmed without losing significant predictive data. In the meantime several deep autoencoders, such as CNN-based AEs, Stacked AEs, Variational AEs (VAE), and LSTM AEs obtain compact representations of the high-dimensional traffic data. The latent features of each autoencoder are then fed into XGBoost, which performs very efficient classification. These results, obtained empirically on several disparate subsets of features, demonstrated consistently better predictive capacity for the hybrid AE + XGBoost models over standalone ensembles in the TII-SSRC-23 dataset. The proposed system is robust for accurately capturing malicious behaviors, and scalable in terms of computation, demonstrating the potential for next-generation IDS solutions through deep autoencoders, NBA-based feature engineering, and gradient-boosted trees.

Keywords Intrusion detection systems · CNN · Stacked · VAE · XGBoost · Feature selection · Network behavior analysis · Malicious traffic detection

---

Present Address:
M. Janati (✉)
National School of Applied Sciences, Sidi Mohamed Ben Abdellah University, Fez, Morocco
e-mail: mohammed.janati@usmba.ac.ma

F. Messaoudi
National School of Business and Management, Sidi Mohamed Ben Abdellah University, Fez, Morocco

Fez, Morocco

F. Messaoudi
e-mail: faycal.messaoudi@usmba.ac.ma

M. Azrour and A. Alabdulatif (eds.), *AI-Driven Security for Next-Generation IoT Systems*, https://doi.org/10.1007/978-3-032-08784-3_8

1 Introduction

With the growing penetration of networks and ever-evolving cyber threats, an advancement in Intrusion Detection System (IDS) is a need of the hour to combat malicious activities in real time [1]. Conventional Intrusion Detection System methods such as signature-based and anomaly-based detection mechanisms have been successful against known attacks but fail to detect new, polymorphic, and multi-stage cyberattacks [2]. Consequently, there has been a growing interest in employing machine learning (ML) and deep learning (DL)-based IDS solutions as they are capable of learning complex patterns of network behavior from large datasets, thereby improving detection rates [3, 4].

Earliest IDS studies focused mainly on benchmark data sets, like KDD99 [5] and NSL-KDD 3. Although these datasets offered valuable initial insights into network attack detection, they are no longer sufficient, primarily because they cannot encompass the diversity of contemporary network traffic and attacks [6]. Datasets fairly more recent than these two, for example, CICIDS2017 [7] and UNSW-NB15 [6], attempted to fill the problem space with new attack scenarios. However, these datasets tend to be biased as they include limited benign traffic or imbalanced class distributions. In recognition of these limitations, TII-SSRC-23 dataset [8] has been presented to provide a more thorough coverage of modern benign and malicious traffic in an array of attack scenarios including denial-of-service (DoS), botnets, brute force attacks and advanced persistent threats (APT).

Deep learning techniques such as convolutional neural networks (CNNs) [9] and long short-term memory (LSTM) networks [10] have been established as effective means for network intrusion detection, but they are resource-intensive and are susceptible to overfitting. In contrast, tree-based ensemble methods like XGBoost [11], are very efficient in processing structured data and provide better interpretability. To use the advantages of both paradigms, hybrid deep learning and ensemble-based models have come up as promising IDS solution, fusing deep autoencoders (AEs) for feature extraction and XGBoost for classification.

In this paper, we propose a hybrid IDS framework, which combines the deep autoencoder architectures (CNN-AE, Stacked AE, Variational AE and LSTM AE) with XGBoost, using Network Behavior Analysis (NBA)-inspired feature engineering and chi-squared–based feature selection—as such evolving a low-latency and accurate detection system. Our key contributions are as follows:

- Feature engineering based on NBA is rich, extracting high-level statistical properties of the input and characteristics of the session-level.
- Chi-squared feature selection for dimensionality reduction while retaining significant features
- Hybrid model integration, where deep autoencoders (CNN, Stacked, Variational, and LSTM-based) extract concise features represented the data, and later through XGBoost, classify them.

- Results using TII-SSRC-23 dataset evaluation that outperforms the competition by multiple metrics, including accuracy, precision, recall, F1-score, AUROC, and AUC-PR.

It has been demonstrated experimentally that the novel approach AE + XGBoost outperforms each of the ensemble learning algorithms independently, demonstrating that the combination is an improvement over the traditional. The remainder of this paper is structured as follows: Sect. 2 presents a review of related work in terms of intrusion detection literature; Sect. 3 introduces the proposed system architecture; Sect. 4 presents the experimental results and performance evaluation; finally, Sect. 5 concludes with suggestions for future research.

2 Related Work

Recent advances in Intrusion Detection Systems have increasingly focused on integrating hybrid models combining deep learning (DL) with ensemble learning methods to address the adaptive and sophisticated nature of modern cyber threats [12, 13]. Conventional datasets such as KDD99 [5] and NSL-KDD [14], although foundational, suffered from redundancy and lacked relevance due to outdated attack scenarios [15]. Subsequent datasets including UNSW-NB15 [6] and CICIDS2017 [7]have addressed these issues partially by reflecting contemporary network traffic patterns and attack types, yet continue to experience problems with class imbalance and inadequate representation of advanced persistent threats. The recently developed TII-SSRC-23 dataset [8] has specifically aimed at overcoming these limitations by incorporating realistic benign traffic alongside complex threats like stealthy malware and multi-phase exploits.

Network Behavior Analysis (NBA) [16], emphasizing session-level and flow-based features, has demonstrated improved detection capabilities against encrypted traffic, zero-day exploits, and evasion strategies [17]. Meanwhile, DL techniques such as convolutional neural networks (CNNs) [9], stacked autoencoders (SAE) [18], variational autoencoders (VAE), and long short-term memory (LSTM) networks have effectively captured complex spatial–temporal correlations within network traffic, significantly enhancing IDS performance. However, standalone DL models often face challenges such as high computational cost, reduced interpretability, and susceptibility to adversarial attacks [19].

To overcome these limitations, hybrid IDS models integrating DL with ensemble classifiers like XGBoost have been increasingly proposed, leveraging complementary strengths: the hierarchical feature extraction capabilities of DL and the computational efficiency and interpretability of ensemble models. Recent research demonstrates the substantial advantages of these hybrid approaches. Mirsky et al. [20] combined autoencoders with random forests, achieving robust anomaly detection through latent feature representation. Yin et al. [21] integrated CNN-based autoencoders with XGBoost classifiers, effectively extracting spatial features for enhanced accuracy in anomaly classification tasks. Similarly, Li et al. [19] employed an LSTM-XGBoost hybrid model to effectively handle temporally correlated traffic data, significantly improving detection accuracy for sequential attack patterns. These hybrid methods consistently exhibit greater predictive performance, robustness against varied cyber threats, and improved generalization capability compared to traditional single-model IDS solutions.

Building on these hybrid frameworks, the current research introduces an advanced IDS architecture integrating CNN-AE, Stacked AE, VAE, and LSTM autoencoders with XGBoost, augmented by NBA-inspired feature extraction and chi-squared feature selection techniques. This comprehensive approach, validated on the TII-SSRC-23 dataset, demonstrates superior detection accuracy, robustness, and computational scalability compared to traditional and other hybrid IDS methods.

3 System Architecture

In this chapter, the architecture design of the suggested Intrusion Detection System (IDS) that intends to rely on flow-based analyses, the principles of Network Behavior Analysis (NBA), a chi-squared–based feature selection, and a hybrid machine learning (ML) framework is discussed. The main objective is to maintain strong detection performance while minimizing additional computation time, which is crucial for deployment in resource constrained environments including edge devices or IoT networks [22, 23]. The base data is the TII-SSRC-23 dataset, which contains a variety of benign and malicious traffic profiles.

3.1 Overview of the Proposed IDS Pipeline

Figure 1 depicts the conceptual workflow of the IDS, comprising five major stages:

1. Data Ingestion and Preprocessing
2. NBA Feature Extraction
3. Chi-Squared–Based Feature Selection (EFS-CST)
4. Hybrid ML Module
5. Decision Engine

3.2 Data Ingestion and Preprocessing

Institutions: the system ingests flow records of TII-SSRC-23 [3], a dataset tailor-made for modern threat landscapes. These flow records collect various characteristics of communications, including host and destination ports per communication,

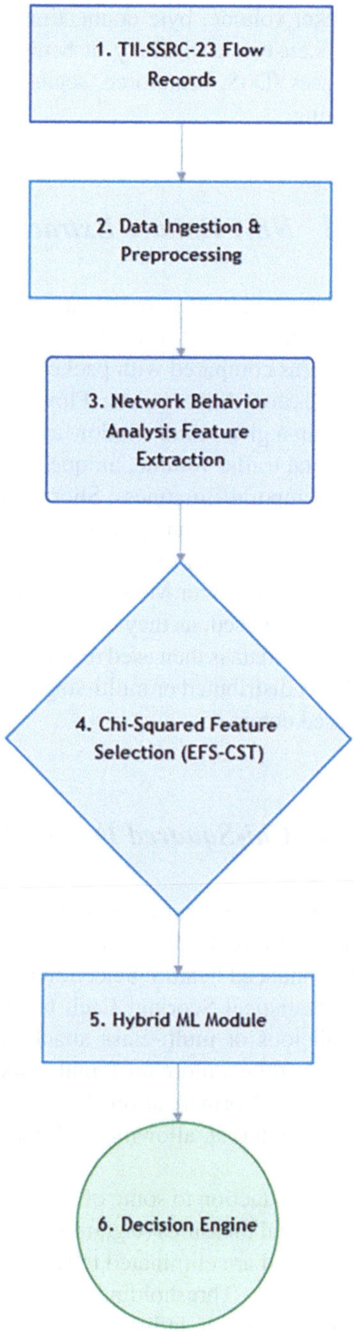

Fig. 1 High-level depiction of the IDS architecture integrating TII-SSRC-23 data ingestion, NBA-inspired feature engineering, chi-squared feature selection, and a hybrid ML module TII-SSRC-23 flow records

packet volume, byte count, time, and TCP-specific (e.g., initial window) metrics. This ensures a diversity of benign (web browsing, video streaming) and malicious classes (DoS, bruteforce, scanning, Mirai botnet) for the use in both training and testing.

3.3 NBA Feature Extraction

Higher-level behavioral features are derived from the preprocessed flows to characterize the NBA [16]. NBA methods focus on aggregated metrics and temporal patterns compared with packet-centric analyses:

Session Aggregation: Flow records matching a given source–destination pair within a given time window are combined, resulting in more granular metrics, such as total traffic volume, unique ports, or session-based duration.

Temporal/Burstiness: Short-term moving averages or exponential weighted averages of flow counts, packet rates, and inter-arrival times are computed to detect sudden spikes (possible DoS) or suspicious dips (stealthy exfiltration).

Rare Behavior Metrics: This feature can identify if unfamiliar protocols or ports are being used, as they can signal scanning or infiltration attempts.

Your data is then used to enrich the IDS, which enhances its ability, for example, to detect distributed or multi-stage attacks that are otherwise transparent to a signature-based detection.

3.4 Chi-Squared Feature Selection (EFS-CST)

Then, we perform a chi-squared test on the remaining features which include 75 original flow-level features and the new NBA features. This procedure, also known as Enhanced Feature Selection using Chi-Squared Test (EFS-CST), is performed as:

Statistical Scoring: Each feature is scored against the target labels (benign vs. malicious or multi-class attack types) resulting in a chi-squared statistic. Higher values reflect more profound association with the outcome variable.

Score Normalization: The raw chi-squared values are linearly transformed into a “0, 1” interval, allowing each feature to be comparable and reducing the impact of outliers.

Introduction to some of the threshold-based methods: A cutoff is decided (trough statistical measures (e.g., min (mean, median normalized scores)) Feature under this threshold are eliminated to reduce the dimensionality and cut noise.

Feature Thresholding: The features with weights greater than the threshold compose a rich, influential set that enables trade-offs between predictive power and computational demands.

EFS-CST helps to answer one of the most enduring challenges in the field of IDS, reducing the high-dimensional input data without sacrificing performance in classifying attacks by discarding low-informative or redundant features.

3.5 *Hybrid DL Module*

This feature subset is then applied to the hybrid machine learning module that includes tree-based algorithms and deep learning models. This dual setup leverages complementary strengths:

Tree-Based Component: Random Forest or XGBoost: Structured data is well-suited for both models and therefore work effectively, even with complex, non-linear relationships among features [11]. Often, these models rank among the top performers on classical intrusion detection benchmarks.

Deep Learning Component: Autoencoders: They are unsupervised neural networks trained on normal flow patterns for reconstruction purposes, so anomalies can be detected by inspecting reconstruction errors [11]. Recurrent Networks (LSTM/GRU): Useful for capturing sequential dependencies in flows, such as for multi-stage or time-dependent attacks (e.g., slow brute force) [5].

An ensemble blends predictions from both component systems through majority voting, weighted averaging and a learned meta-classifier, thereby mitigating the risk of overfitting while improving generalization to novel attacks.

3.6 *Decision Engine*

After classifying every flow record as benign or malicious, it sends the prediction to the decision engine:

Real-Time Deployment: Alerts are triggered quite instantly in risk areas (for instance, potential DoS). Common automated actions include IP blacklisting, rate limiting, or rerouting traffic for further inspection, such as IP blacklisting, rate limiting, or routing potentially suspicious traffic.

- Batch or Forensic Mode: In offline environments where flagged flows are analyzed by a security analyst. Having such an environment allows detailed postmortem analyses to be conducted and the fine-tuning of IDS thresholds to take place.
- Logs and Audit Trails: Every classification event is stored with relevant metadata. This logging capability aids in auditing, false-positive investigations and allows for iteration on models long after deployment.

In this manner, this chapter showed the complete architecture of IDS that proceeds from the TII-SSRC-23 [3] dataset to NBA features and chi-squared-based feature selection in order to generate a thin but powerful predictor set. The system uses ensemble methods (Random Forest or XGBoost) in conjunction with deep learning (autoencoders, RNNs) to accurately identify malicious behaviors ranging from volumetric DoS to stealthy infiltration attempts. The decision engine also supports real-time or batch-mode deployment, with rich logging and audit for analysis. Following that, we test this architecture for performance, resource usage, and generalizability across a variety of experimental scenarios in later chapters.

4 Results and Discussion

4.1 *Experimental Setup*

Experiments were carried on a virtual machine (VM) with a configuration of 126 GB RAM, 12 CPUs and 512 GB of storage. The proposed hybrid IDS framework was evaluated using TII-SSRC-23 dataset 7, which consisted of diverse benign and malicious network traffic. Four variants of autoencoders—CNN-AE, Stacked AE, Variational AE (VAE), and LSTM AE, were combined with an XGBoost classifier separately to detect an attack. Train the autoencoders for 10 epochs with sub_size = 32 and set the number of estimators in the XGboost model to 50. Performance metrics were accuracy, precision, recall, F1-score, AUROC, and AUC-PR.

4.2 *Performance Evaluation Without NBA Features and Chi2 Feature Selection*

No NBA feature extraction or Chi-squared feature selection was used for initial evaluations. The results indicate:

- CNN-AE, Stacked AE, and VAE models consistently approached high accuracy (between 0.98 and 1.00), indicating the models' natural capability to manage raw network data.
- The LSTM AE model performance was much more variable, ranging from accuracy around 0.73 in low feature counts to 0.92–1.00 in higher feature counts, noted since LSTM AE was sensitive on how the features were represented.

4.3 *Performance Evaluation with NBA Features and Chi2 Feature Selection*

The evaluations after the integration of NBA feature extraction and chi-squared (Chi2) feature selection denote evident improvements on model performance as shown on Fig. 2:

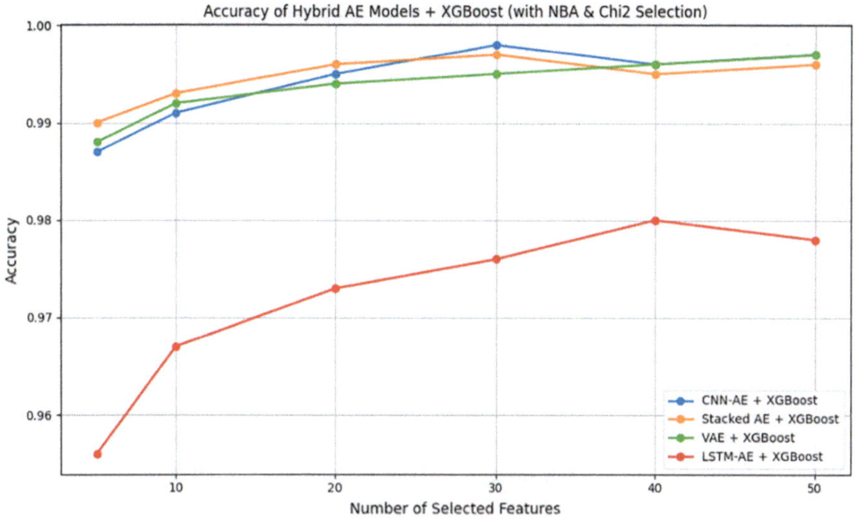

Fig. 2 Comparative accuracy of hybrid autoencoder models combined with XGBoost using NBA features extraction and chi-squared feature selection

- CNN-AE, Stacked AE, and VAE achieve superior and robust performances, revealing accuracies from 0.98 to 1.00 across multiple feature counts, throughout all relevant feature counts further confirming that deriving features from NBA aspects and Chi2 feature extraction is an effective approach.
- The LSTM AE two stages achieved a spectacular improvement of predictive ability preserving a stable high score ranging from 0.95 to 0.99, and it resonates with the positive contribution of NBA features and targeted feature selection methods.

Comparative Analysis

Comparative performance of four hybrid models—CNN-AE, Stacked AE, VAE and LSTM AE—each of which was combined with XGBoost, presented alongside integration of NBA-derived features and chi-squared (Chi2) feature selection. As demonstrated above Table 1, the model performance accuracy (both precision and recall, as well as F1-score and AUC-ROC/AUC-PR) of NBA feature extraction + chi-squared feature selection has the most significant improvement. CNN-AE, Stacked AE, and VAE models stand out among ensemble models for consistently high performance on all metrics, demonstrating robust model characteristics. In contrast, additional stability and predictive performance improvements arise for LSTM AE when using these techniques, underlining the importance of feature selection methods specifically designed for sequential data to fully benefit from their sequencing capabilities.

The enrichment of the input with Chi-squared NBA features extraction and feature selection consistently increased the correctness of the model; particularly, the addition of these features largely improved performance on the LSTM AE model. This

Table 1 Performance comparison of hybrid AE-XGBoost models with and without NBA features and chi-squared selection

Model	Metrics	With NBA/Chi2	Without NBA/Chi2
CNN-AE + XGBoost	Accuracy	1.0000	0.987
	Prec	1.0000	0.987
	Rec	0.9879	0.982
	F1	0.9939	0.9845
	AUC-ROC	0.9886	0.985
	AUC-PR	1.0000	0.990
Stacked AE + XGBoost	Acc	0.9998	0.989
	Prec	1.0000	0.988
	Rec	0.9939	0.986
	F1	0.9969	0.987
	AUC-ROC	0.9930	0.989
	AUC-PR	0.9998	0.989
VAE + XGBoost	Acc	0.9998	0.990
	Prec	0.9999	0.988
	Rec	0.8939	0.980
	F1	*0.9406*	*0.984*
	AUC-ROC	*0.8930*	*0.982*
	AUC-PR	*0.9998*	*0.987*
LSTM + XGBoost	*Acc*	*0.9989*	*0.956*
	Prec	*0.4995*	*0.930*
	Rec	*0.5000*	*0.910*
	F1	*0.6667*	*0.920*
	AUC-ROC	*0.7500*	*0.940*
	AUC-PR	*1.0000*	*0.960*

emphasizes the need for an effective form of feature engineering to maximize both the efficiency and efficacy of the intrusion detection models.

With the help of NBA-derived features and chi-squared based feature selection, the hybrid AE-XGBoost method shows a significant potential to be applied to real-world application of intrusion detection. CNN-AE, Stacked AE and VAE models were all highly robust across all experimental condition. The enhanced stability of LSTM AE also further emphasizes the need and efficacy of customized feature selection for temporal models.

This framework can be effectively applied in a practical environment, thanks to the combination of deep learning-based feature extraction and the ensemble classifiers that address contemporary cybersecurity challenges.

5 Conclusion

This study proposed a hybrid model to audit an intrusion detection framework based on CNN-AE, Stacked AE, VAE, and LSTM AE architectures by embedding XGBoost, as well as Network Behavior Analysis and chi-squared feature extraction. The TII-SSRC-23 dataset evaluation showcased very satisfactory performance, especially with NBA features and Chi2 feature selection. While CNN-AE, Stacked AE, and VAE models showed continuously good results, LSTM AE exhibited significant improvement in stability through focused feature selection.

The proposed framework has several advantages in terms of computational efficiency, scalability, and accuracy uniquely positioned for real-time network security applications. Future research directions are discussed including the need for improved temporal modeling techniques, adaptive feature selection approaches and practical validation of the proposed models.

References

1. H. Assmi *et al.*, "A Robust Security Detection Strategy for Next Generation IoT Networks," *CMC*, vol. 82, no. 1, pp. 443–466, 2025, https://doi.org/10.32604/cmc.2024.059047.
2. K. Bella *et al.*, "An efficient intrusion detection system for IoT security using CNN decision forest," *PeerJ Computer Science*, vol. 10, p. e2290, Sep. 2024, https://doi.org/10.7717/peerj-cs.2290.
3. A. L. Buczak and E. Guven, "A survey of data mining and machine learning methods for cyber security intrusion detection," *IEEE Communications surveys & tutorials*, vol. 18, no. 2, pp. 1153–1176, 2015.
4. A. Blali, S. Dargaoui, M. Azrour, A. Guezzaz, F. Amounas, and A. Alabdulatif, "Analysis of deep learning-based intrusion detection systems in IoT environments," *EDPACS*, pp. 1–35, 2025.
5. A. M. Al Tobi and I. Duncan, "KDD 1999 generation faults: A review and analysis," *Journal of Cyber Security Technology*, vol. 2, no. 3–4, pp. 164–200, 2018.
6. N. Moustafa and J. Slay, "UNSW-NB15: a comprehensive data set for network intrusion detection systems (UNSW-NB15 network data set)," in *2015 military communications and information systems conference (MilCIS)*, IEEE, 2015, pp. 1–6.
7. I. Sharafaldin, A. H. Lashkari, and A. A. Ghorbani, "Toward generating a new intrusion detection dataset and intrusion traffic characterization.," *ICISSp*, vol. 1, no. 2018, pp. 108–116, 2018.
8. D. Herzalla, W. T. Lunardi, and M. Andreoni, "TII-SSRC-23 dataset: typological exploration of diverse traffic patterns for intrusion detection," *IEEE Access*, vol. 11, pp. 118577–118594, 2023.
9. D. Chamou *et al.*, "Intrusion detection system based on network traffic using deep neural networks," in *2019 IEEE 24th international workshop on computer aided modeling and design of communication links and networks (CAMAD)*, IEEE, 2019, pp. 1–6.
10. F. Laghrissi, S. Douzi, K. Douzi, and B. Hssina, "Intrusion detection systems using long short-term memory (LSTM)," *Journal of Big Data*, vol. 8, no. 1, p. 65, 2021.
11. T. Chen and C. Guestrin, "Xgboost: A scalable tree boosting system," in *Proceedings of the 22nd acm sigkdd international conference on knowledge discovery and data mining*, 2016, pp. 785–794.

12. W. Shafik and M. Azrour, "Building a Greener World: Harnessing the Power of IoT and Smart Devices for Sustainable Environment," in *Technical and Technological Solutions Towards a Sustainable Society and Circular Economy*, J. Mabrouki and A. Mourade, Eds., Cham: Springer Nature Switzerland, 2024, pp. 35–58. https://doi.org/10.1007/978-3-031-56292-1_3.
13. M. Mohy-eddine, A. Guezzaz, S. Benkirane, and M. Azrour, "An efficient network intrusion detection model for IoT security using K-NN classifier and feature selection," *Multimedia Tools and Applications*, 2023, https://doi.org/10.1007/s11042-023-14795-2.
14. H. Yzzogh and H. Benaboud, "Enhancing SDN security with a feature-based approach using multiple k-means, Word2Vec, and neural network," *Bulletin of Electrical Engineering and Informatics*, vol. 14, no. 2, pp. 1456–1467, 2025.
15. D. D. Protić, "Review of KDD Cup '99, NSL-KDD and Kyoto 2006+ datasets," *Vojnotehnički glasnik/Military Technical Courier*, vol. 66, no. 3, pp. 580–596, 2018.
16. K. Xu, *Network Behavior Analysis*. Springer, 2022.
17. H. Hindy, R. Atkinson, C. Tachtatzis, J.-N. Colin, E. Bayne, and X. Bellekens, "Utilising deep learning techniques for effective zero-day attack detection," *Electronics*, vol. 9, no. 10, p. 1684, 2020.
18. B. A. Manjunatha, K. A. Shastry, E. Naresh, P. K. Pareek, and K. T. Reddy, "A network intrusion detection framework on sparse deep denoising auto-encoder for dimensionality reduction," *Soft Computing*, vol. 28, no. 5, pp. 4503–4517, 2024.
19. X. Li, W. Chen, Q. Zhang, and L. Wu, "Building auto-encoder intrusion detection system based on random forest feature selection," *Computers & Security*, vol. 95, p. 101851, 2020.
20. Y. Mirsky, T. Tshman, Y. Elovici, and A. Shabtai, "Kitsune: an ensemble of autoencoders for online network intrusion detection," *arXiv preprint* arXiv:1802.09089, 2018.
21. C. Yin, Y. Zhu, J. Fei, and X. He, "A deep learning approach for intrusion detection using recurrent neural networks," *Ieee Access*, vol. 5, pp. 21954–21961, 2017.
22. K. Pradeep Mohan Kumar, M. Saravanan, M. Thenmozhi, and K. Vijayakumar, "Intrusion detection system based on GA-fuzzy classifier for detecting malicious attacks," *Concurrency and Computation: Practice and Experience*, vol. 33, no. 3, p. e5242, 2021.
23. C. Rajathi and P. Rukmani, "Hybrid Learning Model for intrusion detection system: A combination of parametric and non-parametric classifiers," *Alexandria Engineering Journal*, vol. 112, pp. 384–396, 2025.

Ensemble Learning Strategies for Anomaly-Based Intrusion Detection in IoMT Systems

Mohammed Yacoubi, Omar Moussaoui, and Cyril Drocourt

Abstract Intrusion Detection Systems (IDS) are vital components in ensuring the security of intelligent and connected environments, particularly in domains where data sensitivity and system reliability are critical, such as the Internet of Medical Things (IoMT). In this study, we propose an anomaly-based IDS powered by ensemble machine learning techniques to detect and classify cyber threats effectively. The heterogeneity and high vulnerability of IoMT systems demand advanced detection mechanisms capable of identifying various attack patterns in real time. We evaluate five supervised machine learning models Random Forest, CatBoost, LightGBM, XGBoost, and a Stacking ensemble using the CICIoMT2024 dataset, which encompasses a wide range of attack types grouped into six categories: DDoS, DoS, MQTT-based attacks, reconnaissance, spoofing, and benign traffic. Our experimental results demonstrate that the proposed ensemble-based IDS achieves a maximum accuracy of 99.39%, with strong performance across other evaluation metrics. These findings highlight the robustness and scalability of our approach in enhancing anomaly detection and protecting sensitive data within IoMT infrastructures.

Keywords IDS · Cybersecurity · IoMT · Machine learning · CICIOMT2024

M. Yacoubi (✉) · O. Moussaoui
MATSI Laboratory, ESTO, Mohammed First University, Oujda, Morocco
e-mail: mohammed.yacoubi@ump.ac.ma

O. Moussaoui
e-mail: o.moussaoui@ump.ac.ma

C. Drocourt
MIS, UPJV, Amiens, France
e-mail: cyril.drocourt@u-picardie.fr

M. Azrour and A. Alabdulatif (eds.), *AI-Driven Security for Next-Generation IoT Systems*, https://doi.org/10.1007/978-3-032-08784-3_9

1 Introduction

The IoMT is transforming healthcare by enabling continuous monitoring and immediate access to data [1]. However, the expanding network of connected devices makes medical equipment vulnerable to significant cybersecurity threats, risking the confidentiality, integrity, and availability (CIA) of sensitive information and potentially compromising patient safety. Confidentiality ensures data access is restricted, integrity maintains data accuracy, and availability ensures timely access to critical systems [2]. Together, these principles are essential for securing IoMT devices and protecting patient safety. The IoMT architecture is typically structured into four layers [3], each with specific security risks. The perception layer includes physical devices (sensors, medical equipment, etc.) vulnerable to physical attacks and data interception. The network layer enables communication between these devices and central systems, exposing it to attacks like MITM and DDoS. The middleware layer processes and manages data, posing risks of data corruption, unauthorized access, or malware infiltration. Lastly, the application layer handles software for managing and analyzing health data, susceptible to threats such as phishing and data theft. To enhance security, AI techniques integrated with IDS are used to efficiently detect abnormal behaviors and protect IoMT devices and patient data [4]. This article explores using machine learning techniques, including XGBoost, LightGBM, and CatBoost, to develop an IDS for IoMT devices. The goal is to enhance security against cyber threats, particularly during the COVID-19 pandemic, when IoMT devices are crucial for remote patient monitoring and safeguarding sensitive data [5]. Researchers have used deep learning models such as MLP, Convolutional Neural Networks (CNN), Long Short-Term Memory (LSTM), and CNN-LSTM to enhance security threat detection in IoMT environments. Traditional security methods often struggle with the growing complexity and scale of medical IoT networks, leading to high false positive rates and inefficiencies in anomaly detection. These methods may not be agile enough to keep up with the evolving threat landscape in digital healthcare. To overcome these challenges, we propose an IDS based on Gradient Boosting Machines techniques, enhanced with a bagging module and ensemble machine learning methods. This approach aims to improve detection accuracy and minimize false positives, addressing the increasing need for robust security in IoMT devices.

The structure of the paper is as follows: Section 2 presents the necessary background and an overview of related studies. Section 3 is dedicated to the analysis and discussion of the results before concluding the paper.

2 Backgrounds and Related Works

2.1 Internet of Medical Things

IoMT refers to a specialized branch of the Internet of Things (IoT) designed for healthcare applications. It integrates medical devices to enhance patient monitoring, diagnosis, and treatment [6]. By enabling real-time data access and continuous monitoring, IoMT improves patient care and healthcare efficiency. However, its resource-limited devices are vulnerable to various cyber threats, such as Denial of Service (DoS), Distributed Denial of Service (DDoS), and Man-in-the-Middle (MitM) attacks, making intrusion detection a critical challenge to ensure the security and stability of medical services.

To understand IoMT system operations, examining its core architecture is essential, as it forms the backbone for real-time medical data collection, transmission, and decision-making. As shown in Fig. 1, the main components of a typical IoMT system include sensors, gateways, servers or cloud infrastructure, monitoring devices, and healthcare staff, all of which interact in a coordinated manner to support continuous and remote healthcare delivery. Sensors are embedded in various wearable devices (such as smartwatches or fitness bands), implantable medical devices (like pacemakers or glucose monitors), or integrated into traditional medical equipment (such as ECG or blood pressure monitors). These devices acquire live physiological measurements from patients, such as cardiac rhythm, arterial pressure, core temperature, breathing rate, and blood oxygen saturation. The ability to gather data continuously and unobtrusively allows for proactive monitoring and early detection of anomalies [7]. Once data is collected, it is transmitted to the next layer through IoMT gateways. These gateways act as intermediaries between edge devices and central systems. They handle data aggregation, protocol translation, initial filtering or compression, and most importantly, secure data transmission using encryption protocols (e.g., TLS or VPNs). By reducing latency and minimizing data loss, gateways ensure reliable communication even in resource-constrained environments [8]. The servers or cloud platforms form the computational core of the IoMT architecture. They are responsible for storing large volumes of patient data, performing data preprocessing, and applying advanced analytics or machine learning models to generate clinical insights. These platforms can detect patterns, raise alerts in case of abnormal values, and assist in generating predictive models for chronic disease management or emergency risk assessment [9]. Monitoring devices, such as healthcare dashboards, mobile applications, or tablet interfaces, provide a visual representation of the collected and processed health data. These interfaces are designed to be user-friendly and offer functionalities such as real-time charting, trend analysis, and automatic notifications, enabling medical professionals to assess a patient's condition at a glance. Finally, healthcare staff (doctors, nurses, and medical technicians) use these insights to track patient health, respond to critical alerts, adjust treatments, and deliver personalized and timely medical care. The integration of automated monitoring and human expertise enhances the efficiency and effectiveness of

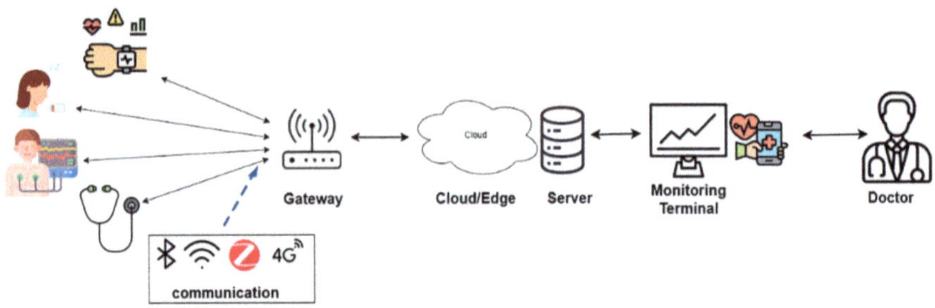

Fig. 1 Architecture of an IoMT network

healthcare, reduces the burden on hospital resources, and supports remote or home-based care, which is especially important for patients with mobility constraints or chronic conditions [10].

2.2 *Security Threats to IoMT*

The swift expansion of IoMT devices has brought forth substantial security challenges, increasing their exposure to diverse cyber threats that may endanger patient safety, undermine data integrity, and disrupt the continuity of healthcare services. Among the various cyber threats, DoS and DDoS attacks pose serious risks by overwhelming IoMT systems with traffic, rendering critical devices or services inaccessible and disrupting healthcare operations [11]. MitM attacks intercept communications between devices or servers, compromising the Confidentiality and Integrity of patient data by allowing attackers to eavesdrop, alter, or inject malicious information into data exchanges [12]. Similarly, spoofing attacks, where attackers impersonate legitimate devices or users, can lead to unauthorized access and data manipulation, undermining both Integrity and Confidentiality [11]. Reconnaissance attacks, while often the precursor to more damaging actions, involve attackers gathering information about IoMT systems, potentially identifying vulnerabilities that could be exploited in future attacks, thus affecting all three principles of the CIA triad. Malware, which encompasses various forms of malicious software, can infect IoMT devices, corrupting data, and disrupting services, compromising Integrity, Availability, and Confidentiality [13]. Lastly, Ransomware attacks can lock critical healthcare data or devices, demanding a ransom for access, severely affecting the Availability of services and potentially damaging the Integrity of the data [14]. These diverse attacks highlight the importance of securing IoMT systems to safeguard patient data, ensure service continuity, and protect the overall functionality of healthcare networks.

2.3 Intrusion Detection System (IDS)

An IDS is a software solution that monitors the behavior of IoT devices to detect malicious activities. IDS typically relies on three main detection strategies: anomaly-based detection, which flags deviations from established normal behavior; signature-based detection, which compares current activities to a repository of known attack patterns; and hybrid detection, which combines both methods to enhance accuracy and minimize false positives. IDS can be categorized into three main types: network-based IDS (NIDS), host-based IDS (HIDS), and hybrid IDS [15]. Network-based IDS (NIDS) monitor traffic across the network in real time, inspecting packets as they traverse communication channels to detect suspicious patterns or known attack signatures. This enables early detection of threats such as DDoS attacks, scanning activities, or data exfiltration attempts before they reach critical endpoints. Host-based IDS (HIDS), on the other hand, are installed directly on individual devices or endpoints. They analyze logs, file integrity, system calls, and other host-specific activities to detect anomalies or security breaches. HIDS are particularly effective in identifying insider threats, unauthorized modifications, or malware infections that may not be visible at the network level. Finally, hybrid IDS combine the capabilities of both NIDS and HIDS, leveraging their respective strengths to provide a comprehensive, multi-layered defense mechanism. By correlating data from both network and host sources, hybrid systems enhance detection accuracy, reduce false positives, and provide greater contextual awareness an essential feature for securing complex and dynamic IoMT infrastructures. In this context, the integration of artificial intelligence (AI) into the construction process of IDS, as illustrated in Fig. 2, plays a critical role in detecting zero-day attacks previously unknown threats that exploit undiscovered vulnerabilities. Unlike signature-based systems, which rely on predefined patterns, AI-driven IDS leverage machine learning to identify abnormal behaviors, enabling proactive detection of novel threats and enhancing the overall resilience of IoMT systems.

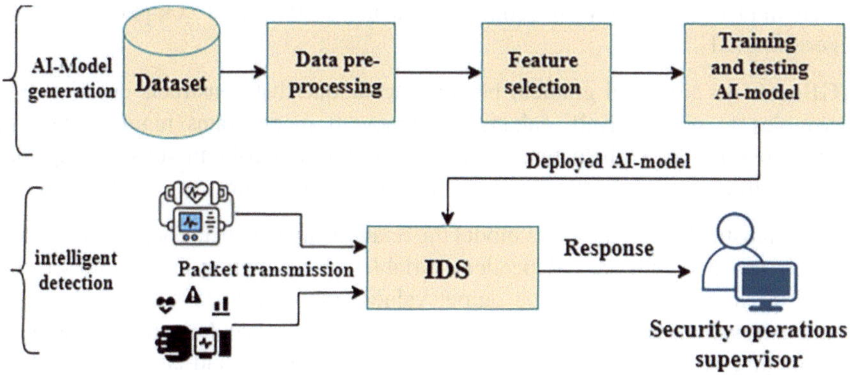

Fig. 2 Process of building an AI-driven IDS

2.4 Machine Learning Algorithms

In our research, we focused on supervised learning, particularly in addressing classification problems [16], which is crucial in the field of cybersecurity for connected medical devices. These methods involve training models on labeled datasets to enable accurate prediction of predefined categories such as distinguishing between normal and malicious network traffic. This is especially critical in healthcare environments where the integrity, confidentiality, and availability of sensitive data must be ensured. In this context, ML algorithms are widely utilized for IDS, allowing the identification and mitigation of various cyber threats by analyzing patterns in network traffic data [17]. Our goal is to enhance the security of connected medical devices by using classification models, incorporating techniques like boosting, bagging, and stacking modules, including classifiers such as Random Forest (RF), CatBoost, and Logistic Regression (LR), to prevent cyberattacks and strengthen device protection [18].

RF: A bagging ensemble algorithm that builds multiple decision trees, each trained on random subsets of the training data (with replacement) and a random selection of features at each split. This randomness introduces diversity among the trees, which reduces the correlation between them. The collective predictions of these trees are combined through majority voting for classification or averaging for regression. This ensemble approach helps to reduce overfitting, improve generalization, and enhance the model's stability and accuracy, especially in high-dimensional and noisy datasets [19].

CatBoost: A gradient boosting algorithm designed to efficiently handle categorical data, automatically processing non-numeric variables. It utilizes ordered boosting to minimize errors and accelerate training, making it suitable for fast and accurate predictions on complex datasets. CatBoost also offers GPU support for improved performance [20].

LightGBM: A fast, distributed gradient boosting framework optimized for low memory usage and large datasets. It employs a leaf-wise tree splitting approach, which improves accuracy by reducing loss more effectively compared to other algorithms [21].

XGBoost: An advanced gradient boosting technique that constructs decision tree ensembles to progressively enhance accuracy. It incorporates regularization to prevent overfitting, with features like tree pruning and column subsampling that improve efficiency while minimizing computation and memory requirements [22].

LR: A statistical method used to model the relationship between a dependent binary variable and one or more independent variables by estimating probabilities using a logistic sigmoid function. The output values are constrained between 0 and 1, making it well-suited for binary classification tasks. It calculates the log-odds of the outcome as a linear combination of the input features, and applies maximum likelihood estimation to optimize the model parameters [23].

2.5 Ensemble Learning Techniques

Ensemble learning techniques combine multiple base models with the aim of improving detection performance, robustness, and generalization of IDS in IoMT environments [24]. These techniques are generally divided into three main categories: bagging, boosting, and stacking, as illustrated in Fig. 3.

- **Bagging (Bootstrap Aggregating)**: As illustrated on the left side of the figure, the bagging technique constructs several datasets (D1, D2, D3) by randomly selecting instances from the original dataset D, allowing duplicates. Each of these datasets is then used to train a separate model m, most commonly a decision tree. To produce the final result, the outputs of all models are merged, most often by using majority vote or by calculating the average. In our study, we utilized the RF algorithm, which builds a large number of decision trees on different bootstrap samples and feature subsets, then combines their outputs to improve overall accuracy and reduce overfitting.
- **Boosting**: In contrast to bagging, boosting builds models sequentially, where each new model is trained to correct the errors made by the previous ones. As depicted in the middle of Fig. 3, the first model mmm is trained on the full dataset D. The performance of this model is then evaluated, and greater weight is assigned to the instances it misclassified. This reweighted dataset is used to train the next model, which focuses more on the difficult samples. This process continues iteratively, forming a sequence of models where each learner tries to reduce the errors of its predecessor. The final prediction is obtained through a weighted aggregation of the outputs of all models. In our study, we employed CatBoost, LightGBM, and XGBoost, three efficient gradient boosting algorithms known for their strong predictive performance, robustness, and scalability on structured data.
- **Stacking**: The right part of the figure illustrates the stacking approach, where multiple diverse base learners (e.g., model 1, model 2, model 3) are trained in

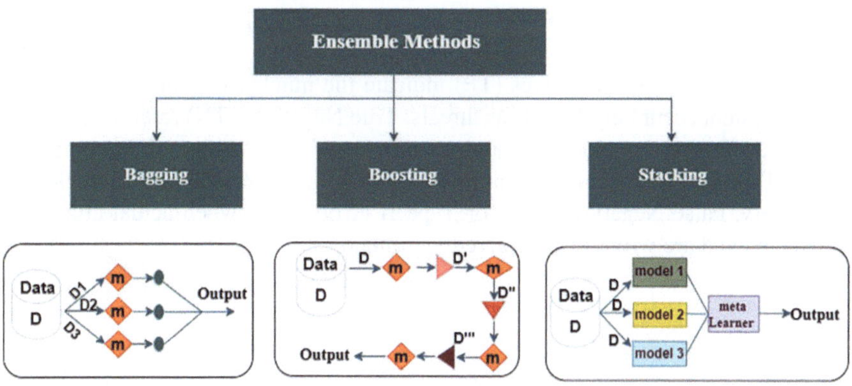

Fig. 3 Overview of ensemble learning methods: bagging, boosting, and stacking

parallel on the same dataset D. Their predictions are then used as input features for a meta-learner, which learns to combine these predictions into a final decision. In our proposed model, we used RF and CatBoost as base-level models. The outputs of these models were passed to a LR classifier acting as the meta-learner. This architecture benefits from the diversity of the base models and the ability of the meta-learner to exploit their strengths, resulting in enhanced generalization and detection capability.

2.6 Evaluation Metrics for Classification Models

To objectively evaluate the performance of classification models in IDS, several standard metrics are utilized, including accuracy, precision, recall, and F1-score [25]. Each of these metrics offers a different perspective on the model's ability to distinguish between normal and malicious behaviors. Accuracy reflects the overall correctness of the predictions, while precision indicates the proportion of true positives among all predicted positives. Recall (or sensitivity) measures the model's ability to detect actual positive cases, and the F1-score provides a harmonic mean between precision and recall, which is particularly valuable in scenarios with class imbalance.

$$\text{Accuracy} = \frac{\text{TP} + \text{TN}}{\text{TP} + \text{TN} + \text{FP} + \text{FN}} \tag{1}$$

$$\text{Precision} = \frac{\text{TP}}{\text{TP} + \text{FP}} \tag{2}$$

$$\text{Recall} = \frac{\text{TP}}{\text{TP} + \text{FN}} \tag{3}$$

$$\text{F1} - \text{score} = 2\frac{\text{Precision.Recall}}{\text{Precision} + \text{Recall}} \tag{4}$$

In this context, True Positives (TP) indicate the number of malicious activities that the system accurately detects as threats. True Negatives (TN) refer to legitimate network traffic that is correctly recognized as benign. False Positives (FP), commonly termed Type I errors, occur when normal traffic is mistakenly flagged as malicious. Conversely, False Negatives (FN), or Type II errors, arise when actual attacks go undetected and are wrongly considered legitimate.

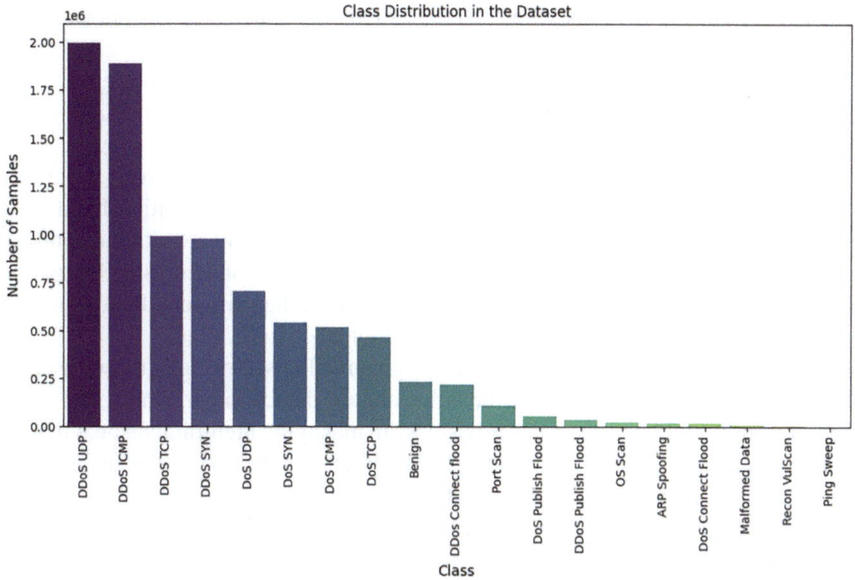

Fig. 4 Class distribution with the CICIoMT2024 attack dataset

2.7 Dataset

Designed specifically for the healthcare context, the CICIoMT2024 dataset [26] provides a valuable foundation for analyzing the security of IoMT systems. It originates from a testbed of 40 devices (25 physical, 15 virtual) simulating real-world attack scenarios, covering protocols like BLE, Wi-Fi, and MQTT. Available in CSV format, the dataset supports binary, multiclass (six categories), and 19-type attack classification [27]. These attacks include ARP Spoofing, Ping Sweep, Port Scan, VulScan, and various DoS/DDoS methods (ICMP, SYN, TCP, UDP, Connect Flood, Publish Flood), along with Malformed Data. Each record contains 45 features, with instance distribution shown in Fig. 4. Our study focuses on the 19-type attack classification.

2.8 Related Works

This section reviews existing research on IDS in IoMT environments, with a particular focus on the application of artificial AI, especially ML, to enhance cybersecurity in connected medical devices. AI techniques are increasingly being integrated into IDS frameworks due to their ability to learn from data, adapt to evolving attack patterns, and provide high detection accuracy in real-time scenarios a necessity for sensitive healthcare systems where false positives or undetected threats can have critical consequences.

Alalwany et al. [28] emphasize the superiority of ensemble learning methods, particularly highlighting the Stacking classifier, which achieved outstanding results with an accuracy of 0.9559, precision of 0.9558, recall of 0.9559 and an F1-score of 0.9556. These high-performance metrics demonstrate the model's strong capability to effectively differentiate between benign and malicious activity a critical function in healthcare environments where data integrity and patient safety are vital. In contrast, more traditional approaches such as AdaBoost, Naïve Bayes, and Logistic Regression showed comparatively lower performance, indicating their limited effectiveness in addressing the complexity and heterogeneity typical of IoMT data.

Moreover, Zhang et al. [29] show that Bagging methods, especially those based on the Random Forest algorithm, significantly enhance detection performance. Thanks to its ensemble architecture and its robustness in managing high-dimensional feature spaces, Random Forest is particularly well-suited for analyzing the diverse and complex traffic patterns characteristic of IoMT systems.

Saba [30] proposes an Intrusion Detection System (IDS) for smart hospitals using ensemble classifiers combined with Principal Component Analysis (PCA) for feature reduction. The study, evaluated on the KDDCup-99 dataset, compares ensemble methods such as Bagging (bagged decision trees, random forest), Boosting (AdaBoost, stochastic gradient boosting), and Voting Ensemble. Results demonstrate that Bagging particularly bagged decision trees achieves the highest performance with 93.2% accuracy, 99.9% precision, 92% recall, and 96.1% F-score. These robust metrics underscore the efficacy of ensemble approaches in addressing the complexity and heterogeneity of IoMT data, critical for healthcare environments where data security and connected medical device reliability are paramount. In comparison, Boosting and Voting Ensemble methods exhibit slightly lower performance, though still competitive. This work highlights the value of advanced machine learning techniques in strengthening cybersecurity for smart healthcare systems, aligning with contemporary challenges in safeguarding sensitive patient data and mitigating targeted cyberattacks.

Goumidi et al. [31] proposed a real-time anomaly detection model for IoMT, using ensemble learning to address domain-specific cybersecurity challenges. The authors developed a novel medical dataset by combining physiological data (BRFSS) with attack patterns inspired by UNSW-NB15, along with additional healthcare-relevant threats, resulting in improved anomaly detection performance. A stacking ensemble model XGBoost as the meta-learner over Random Forest and ANN achieved 98.02% accuracy and demonstrated robustness in real-time message classification. This work underscores the importance of tailored datasets and ensemble techniques in enhancing IoMT security, while also highlighting the need for broader attack coverage and validation across diverse datasets to improve generalizability.

In our study, we evaluate ensemble-based machine learning models namely bagging, boosting, and stacking on the newly introduced CICIOMT2024 dataset, aiming to assess and compare their effectiveness in detecting diverse and healthcare-specific cyberattacks within IoMT environments.

Table 1 IoMT cyberattack dataset

Dataset	Sample of training	Sample of testing
CICIOTM2024	7,160,831	1,614,182

3 Results and Discussion

3.1 *Hardware and Software Characteristics*

Details of the server's hardware setup used during the experiments are provided below.

- CPU(s): Intel(R) Xeon(R) CPU E5-2640 v4 @ 2.40 GHz, (20cores)
- RAM: 125 GB

The experiments were conducted using Scikit-learn, a free, open-source Python library designed for efficient machine learning applications [32].

3.2 *Evaluating the Result*

This study enhances cyberattack detection in IoMT systems using bagging, boosting, and stacking techniques. The proposed architecture leverages StandardScaler preprocessing to improve performance. Key factors influencing model robustness are analyzed using metrics such as accuracy, precision, and training time, based on the full CICIOTM2024 dataset comprising 19 attack types. Table 1 outlines training and testing sample sizes, while Table 2 details model configurations.

Table 3 summarizes the evaluation results and computational performance of RF model applied to the intrusion detection task. The model achieved a high accuracy

Table 2 Model's details and hyperparameters

Model	Hyperparameters
RF	n_estimators = 100, criterion = "gini", min_saples_split = 2,min_saples_leaf = 1, min_weight_fraction_leaf = 0.0,max_fetures = "sqrt",min_impurity_decrease = 0.0, bootstrap = True, oob_score = False,warm_start = False, ccp_alpha = 0.0, random_state = 42
CatBoost	iterations = 120, depth = 10, learning_rate = 0.1, loss_tion = 'MutClass', verbose = 50, task_type = 'CPU', random_seed = 42
LightGBM	num_leaves = 20, max_depth = -1, learning_rate = 0.01, n_estimators = 200, metric = 'multi_logloss',random_state = 20
XGBoost	objective = 'multi:softprob',num_class = 19,max_depth = 10,'eta' = 0.1, eval_metric = 'mlogloss', tree_method = 'hist'
LR	LogisticRegression()

of 99.39%, indicating that the vast majority of instances both normal and malicious were correctly classified. In terms of precision, the model scored 87.05%, meaning that out of all the instances identified as attacks, a significant portion were indeed actual threats. The recall, measured at 89.86%, reflects the model's strong ability to correctly identify most of the true attack samples in the dataset. These two metrics are further consolidated by the F1-score of 87.56%, which offers a harmonic mean and represents a good trade-off between precision and recall. From a computational standpoint, the model's training time was 12.16 s, while the testing phase required 25.49 s. These durations are considered efficient, especially in the context of resource-constrained environments like IoMT systems, suggesting that the RF model can be deployed in scenarios where quick detection and decision-making are essential.

Table 4 demonstrates a significant improvement in our study, achieving 99.39% accuracy compared to 73.3% in previous work, along with notable enhancements in precision, recall, and F1-score.

Table 5 provides a comparative analysis of three gradient boosting models: CatBoost, LightGBM, and XGBoost, highlighting their trade-offs in terms of accuracy, precision, recall, and computational efficiency. Among the three, CatBoost achieves the highest performance metrics, with an accuracy of 99.36%, precision of 86.10%, and recall of 89.10%. These results indicate that CatBoost is particularly effective in correctly identifying both normal and malicious instances, making it a strong candidate for highly accurate intrusion detection.

However, this high performance comes at a computational cost. CatBoost records the longest training time, reaching 683.67 s, which may be a limiting factor in time-sensitive or resource-constrained environments such as IoMT systems.

LightGBM, in contrast, offers a well-balanced compromise. While its accuracy and recall are slightly lower than CatBoost, it still achieves competitive results with a training time of just 67.79 s, making it more suitable for real-time or near-real-time applications where fast model deployment is critical.

XGBoost ranks slightly lower in terms of detection performance but stands out for its computational efficiency. It records the shortest training time of 55.08 s and

Table 3 Performance evaluation of bagging method (random forest)

Model	Accuracy (%)	Precision (%)	Recall (%)	F1-score (%)	Time training (s)	Time testing (s)
RF	99.39	87.05	89.86	87.56	12.16	25.49

Table 4 Comparison of our study with other studies on intrusion detection using RF on the CICIoMT2024 dataset

Work	Model	Accuracy (%)	Precision (%)	Recall (%)	F1-score (%)
[26]	RF	73.3	69.1	57.7	55.1
Ours		99.39	87.05	89.86	87.56

Table 5 Performance evaluation of boosting methods for three modules

Model	Accuracy (%)	Precision (%)	Recall (%)	F1-score (%)	Time training (s)	Time testing (s)
CatBoost	99.36	86.10	89.10	87.03	683.67	4.50
LightGBM	99.28	83.86	87.16	83.81	67.79	218.23
XGBoost	99.01	82.26	87.72	83.44	55.08	24.56

the fastest testing time of 24.56 s, making it ideal for scenarios that prioritize speed over marginal gains in accuracy.

Table 6 illustrates the structure and performance of the proposed stacking model, which is built upon a two-layer architecture designed to enhance classification accuracy in IoMT intrusion detection. In the first layer, two powerful ensemble methods—CatBoost and RF are used to generate probability estimates for each input sample. These probabilities are then passed to the second layer, where a LR model acts as a meta-classifier, combining the outputs to make the final prediction.

This configuration achieves a high overall accuracy of 99.39%, along with strong performance on other key metrics: 87.29% precision, 89.04% recall, and an F1-score of 86.91%. These results indicate that the model is effective at both identifying true attacks and minimizing false positives and negatives. However, the increased complexity of the stacking model comes at a computational cost. Training the model takes approximately 701.13 s, while testing requires 98.53 s. This is notably higher than the time required for individual classifiers. For comparison, CatBoost and RF used separately yield similar results, but Logistic Regression alone achieves only 74.41% accuracy. This underscores the benefit of combining multiple models: the ensemble approach leverages the strengths of each component to deliver more accurate and robust predictions. The integration of CatBoost and RF as base learners, with LR as a meta-learner, significantly boosts the model's ability to detect a wide range of cyberattacks in IoMT systems. Despite the additional computational cost, the enhanced generalization across different attack types makes this approach a strong candidate for practical deployment in healthcare cybersecurity scenarios.

Table 6 Performance evaluation of stacking method

Model	Accuracy (%)	Precision (%)	Recall (%)	F1-score (%)	Time training (s)	Time testing (s)
Stacking	99.39	87.29	89.04	86.91	701.13	98.53
RF	99.39	87.05	89.86	87.56	12.16	25.49
CatBoost	99.36	86.10	89.10	87.03	683.67	4.50
LR	74.41	61.42	57.82	54.33	119.78	1.52

4 Conclusion

Our evaluation shows that ensemble methods improve IoMT IDS accuracy and reliability. The stacking approach combining CatBoost, RF, and LR achieves top performance but with higher computational cost. Bagging and boosting methods also perform well, balancing performance and efficiency for IoMT cybersecurity.

Future work could explore deep learning models like CNN and LSTM for capturing complex attack patterns. Hybrid models combining traditional and deep learning techniques could enhance performance, while federated learning offers potential for decentralized training, improving privacy and scalability in IoMT networks. These directions are key to advancing real-time, high-performance IDS solutions for medical IoT systems.

References

1. K. Boikanyo, A. M. Zungeru, B. Sigweni, A. Yahya, and C. Lebekwe, “Remote patient monitoring systems: Applications, architecture, and challenges,” *Sci Afr*, vol. 20, p. e01638, Jul. 2023, https://doi.org/10.1016/J.SCIAF.2023.E01638.
2. T. Sasi, A. H. Lashkari, R. Lu, P. Xiong, and S. Iqbal, “A comprehensive survey on IoT attacks: Taxonomy, detection mechanisms and challenges,” *Journal of Information and Intelligence*, vol. 2, no. 6, pp. 455–513, Nov. 2024, https://doi.org/10.1016/J.JIIXD.2023.12.001.
3. C. Huang, J. Wang, S. Wang, and Y. Zhang, “Internet of medical things: A systematic review,” *Neurocomputing*, vol. 557, p. 126719, Nov. 2023, https://doi.org/10.1016/J.NEUCOM.2023.126719.
4. I. Idrissi, M. Boukabous, M. Grari, M. Azizi, and O. Moussaoui, “An Intrusion Detection System Using Machine Learning for Internet of Medical Things,” *Lecture Notes in Electrical Engineering*, vol. 954, pp. 641–649, 2023, https://doi.org/10.1007/978-981-19-6223-3_66.
5. I. Sy, B. Diouf, B. Diouf, A. K. Diop, C. Drocourt, and D. Durand, “AI-driven Threat Management in Healthcare Systems,” *Intelligent Cybersecurity and Resilience for Critical Industries: Challenges and Applications*, no. 1, pp. 7–31, Feb. 2025, https://doi.org/10.1201/9788770047746-3.
6. M. Osama *et al.*, “Internet of Medical Things and Healthcare 4.0: Trends, Requirements, Challenges, and Research Directions,” *Sensors 2023, Vol. 23, Page 7435*, vol. 23, no. 17, p. 7435, Aug. 2023, https://doi.org/10.3390/S23177435.
7. M. L. Hernandez-Jaimes, A. Martinez-Cruz, K. A. Ramírez-Gutiérrez, and C. Feregrino-Uribe, “Artificial intelligence for IoMT security: A review of intrusion detection systems, attacks, datasets and Cloud–Fog–Edge architectures,” *Internet of Things*, vol. 23, p. 100887, Oct. 2023, https://doi.org/10.1016/J.IOT.2023.100887.
8. O. Debauche *et al.*, “RAMi: A New Real-Time Internet of Medical Things Architecture for Elderly Patient Monitoring,” *Information 2022, Vol. 13, Page 423*, vol. 13, no. 9, p. 423, Sep. 2022, https://doi.org/10.3390/INFO13090423.
9. S. F. Ahmed, M. S. Bin Alam, S. Afrin, S. J. Rafa, N. Rafa, and A. H. Gandomi, “Insights into Internet of Medical Things (IoMT): Data fusion, security issues and potential solutions,” *Information Fusion*, vol. 102, p. 102060, Feb. 2024, https://doi.org/10.1016/J.INFFUS.2023.102060.
10. S. Abdulmalek *et al.*, “IoT-Based Healthcare-Monitoring System towards Improving Quality of Life: A Review,” *Healthcare*, vol. 10, no. 10, p. 1993, Oct. 2022, https://doi.org/10.3390/HEALTHCARE10101993.

11. J. P. A. Yaacoub *et al.*, "Securing internet of medical things systems: Limitations, issues and recommendations," *Future Generation Computer Systems*, vol. 105, pp. 581–606, Apr. 2020, https://doi.org/10.1016/J.FUTURE.2019.12.028.
12. N. Sivasankari and S. Kamalakkannan, "Detection and prevention of man-in-the-middle attack in iot network using regression modeling," *Advances in Engineering Software*, vol. 169, p. 103126, Jul. 2022, https://doi.org/10.1016/J.ADVENGSOFT.2022.103126.
13. A. Si-Ahmed, M. A. Al-Garadi, and N. Boustia, "Survey of Machine Learning based intrusion detection methods for Internet of Medical Things," *Appl Soft Comput*, vol. 140, p. 110227, Jun. 2023, https://doi.org/10.1016/J.ASOC.2023.110227.
14. H. T. Neprash *et al.*, "Trends in Ransomware Attacks on US Hospitals, Clinics, and Other Health Care Delivery Organizations, 2016-2021," *JAMA Health Forum*, vol. 3, no. 12, pp. e224873–e224873, Dec. 2022, https://doi.org/10.1001/JAMAHEALTHFORUM.2022.4873.
15. A. Khraisat, I. Gondal, P. Vamplew, and J. Kamruzzaman, "Survey of intrusion detection systems: techniques, datasets and challenges," *Cybersecurity*, vol. 2, no. 1, pp. 1–22, Dec. 2019, https://doi.org/10.1186/S42400-019-0038-7/FIGURES/8.
16. A. Tharwat, "Classification assessment methods," *Applied Computing and Informatics*, vol. 17, no. 1, pp. 168–192, 2018, https://doi.org/10.1016/J.ACI.2018.08.003/FULL/PDF.
17. R. U. Rasool, H. F. Ahmad, W. Rafique, A. Qayyum, and J. Qadir, "Security and privacy of internet of medical things: A contemporary review in the age of surveillance, botnets, and adversarial ML," *Journal of Network and Computer Applications*, vol. 201, p. 103332, May 2022, https://doi.org/10.1016/J.JNCA.2022.103332.
18. Ö. Aslan, S. S. Aktuğ, M. Ozkan-Okay, A. A. Yilmaz, and E. Akin, "A Comprehensive Review of Cyber Security Vulnerabilities, Threats, Attacks, and Solutions," *Electronics 2023, Vol. 12, Page 1333*, vol. 12, no. 6, p. 1333, Mar. 2023, https://doi.org/10.3390/ELECTRONICS12061333.
19. L. Breiman, "Random forests," *Mach Learn*, vol. 45, no. 1, pp. 5–32, Oct. 2001, https://doi.org/10.1023/A:1010933404324.
20. L. Prokhorenkova, G. Gusev, A. Vorobev, A. V. Dorogush, and A. Gulin, "CatBoost: unbiased boosting with categorical features," *Adv Neural Inf Process Syst*, vol. 2018-December, pp. 6638–6648, Jun. 2017, Accessed: Jan. 25, 2025. [Online]. Available: https://arxiv.org/abs/1706.09516v5
21. G. Ke *et al.*, "LightGBM: A Highly Efficient Gradient Boosting Decision Tree," *Adv Neural Inf Process Syst*, vol. 30, 2017, Accessed: Feb. 09, 2025. [Online]. Available: https://github.com/Microsoft/LightGBM.
22. T. Chen and C. Guestrin, "XGBoost: A scalable tree boosting system," *Proceedings of the ACM SIGKDD International Conference on Knowledge Discovery and Data Mining*, vol. 13–17-August-2016, pp. 785–794, Aug. 2016, https://doi.org/10.1145/2939672.2939785/SUPPL_FILE/KDD2016_CHEN_BOOSTING_SYSTEM_01-ACM.MP4.
23. C. Starbuck, "Logistic Regression," *The Fundamentals of People Analytics*, pp. 223–238, 2023, https://doi.org/10.1007/978-3-031-28674-2_12.
24. M. Ibrahim and A. Al-Wadi, "Enhancing IoMT network security using ensemble learning-based intrusion detection systems," *Journal of Engineering Research*, Dec. 2024, https://doi.org/10.1016/J.JER.2024.12.003.
25. S. M. Kasongo and Y. Sun, "Performance Analysis of Intrusion Detection Systems Using a Feature Selection Method on the UNSW-NB15 Dataset," *J Big Data*, vol. 7, no. 1, pp. 1–20, Dec. 2020, https://doi.org/10.1186/S40537-020-00379-6/TABLES/8.
26. S. Dadkhah, E. C. P. Neto, R. Ferreira, R. C. Molokwu, S. Sadeghi, and A. A. Ghorbani, "CICIoMT2024: A benchmark dataset for multi-protocol security assessment in IoMT," *Internet of Things*, vol. 28, p. 101351, Dec. 2024, https://doi.org/10.1016/J.IOT.2024.101351.
27. "IoMT Dataset 2024 | Datasets | Research | Canadian Institute for Cybersecurity | UNB." Accessed: May 12, 2025. [Online]. Available: https://www.unb.ca/cic/datasets/iomt-dataset-2024.html
28. E. Alalwany, B. Alsharif, Y. Alotaibi, A. Alfahaid, I. Mahgoub, and M. Ilyas, "Stacking Ensemble Deep Learning for Real-Time Intrusion Detection in IoMT Environments," *Sensors*, vol. 25, no. 3, p. 624, Jan. 2025, https://doi.org/10.3390/S25030624.

29. Z. Zhang, S. Kong, T. Xiao, and A. Yang, "A Network Intrusion Detection Method Based on Bagging Ensemble," *Symmetry 2024, Vol. 16, Page 850*, vol. 16, no. 7, p. 850, Jul. 2024, https://doi.org/10.3390/SYM16070850.
30. T. Saba, "Intrusion Detection in Smart City Hospitals using Ensemble Classifiers," *Proceedings—International Conference on Developments in eSystems Engineering, DeSE*, vol. 2020-December, pp. 418–422, Dec. 2020, https://doi.org/10.1109/DESE51703.2020.9450247.
31. H. Goumidi and S. Pierre, "Real-Time Anomaly Detection in IoMT Networks Using Stacking model and a Healthcare-Specific Dataset," *IEEE Access*, 2025, https://doi.org/10.1109/ACCESS.2025.3563158.
32. "scikit-learn: machine learning in Python—scikit-learn 1.6.1 documentation." Accessed: May 13, 2025. [Online]. Available: https://scikit-learn.org/stable/#

A Lightweight Semi-Centralized Authentication Scheme for Internet of Drones (IoD) Networks

Souhayla Dargaoui, Mourade Azrour, Ahmad El Allaoui, Fatima Amounas, Abdeslem Blali, Azidine Guezzaz, and Abdulatif Alabdulatif

Abstract All things considered, most drones operate independently and frequently in remote or unreachable areas through networks. Recently, such networks have integrated into our daily lives, gathering new opportunities and facilities. However, new security challenges arise from these integrations. Researchers have developed various authentication protocols over the past ten years to overcome the arising challenges. Nonetheless, most of the protocols provided failed to resist physical assaults. To solve this problem and improve other security characteristics, this chapter suggests a semi-centralized authentication protocol based on PUF for Drone-to-Drone communication. In our protocol, the gateway only generates PUF challenges when direct authentication between drones is achieved using the Computational Diffie–Hellman problem. Our protocol presents a significant improvement in computation overhead, requiring only 7.5 ms. Likewise, it demands an acceptable communication overhead compared to related protocols.

S. Dargaoui (✉) · M. Azrour · A. E. Allaoui · F. Amounas · A. Blali
IMIA Laboratory, MSIA Team, Faculty of Sciences and Techniques, Moulay Ismail University of Meknes, Errachidia, Morocco
e-mail: s.dargaoui@edu.umi.ac.ma

M. Azrour
e-mail: mo.azrour@umi.ac.ma

A. E. Allaoui
e-mail: a.elallaoui@umi.ac.ma

F. Amounas
e-mail: f_amounas@yahoo.fr

A. Blali
e-mail: ab.blali@edu.umi.ac.ma

A. Guezzaz
Higher School of Technology Essaouira, Cadi Ayyad University, Essaouira, Morocco
e-mail: a.guezzaz@uca.ma

A. Alabdulatif
Department of Computer Science, College of Computer, Qassim University, Buraydah, Saudi Arabia
e-mail: ab.alabdulatif@qu.edu.sa

M. Azrour and A. Alabdulatif (eds.), *AI-Driven Security for Next-Generation IoT Systems*, https://doi.org/10.1007/978-3-032-08784-3_10

Keywords Internet of drones · Drone-to-drone communication · Mutual authentication · PUF · IoD security

1 Introduction

Unmanned Aerial Vehicles (UAVs), commonly known as drones, are increasingly deployed in networks called the Internet of Drones that collect and interact with environmental data [1, 2]. Such networks have caught great attention as they may be significantly used across various sectors, including traffic monitoring, real-time object detection, emergency assistance, and military applications [3–5]. As their applications increase, the necessity for confident and reliable connection among drone networks becomes crucial, especially in collaborative and swarm-based use cases [6–10]. Drone-to-drone communication, however, introduces new security challenges due to the insecure channels used to exchange data and the decentralized and often dynamic nature of UAV networks, making them susceptible to a wide range of cyber attacks such as spoofing, man-in-the-middle attacks, and unauthorized access [11–19].

On the others hands, current authentication protocols often combine multiple factors, referred to as multi-factor authentication, to enhance security by reducing the risk associated with credential compromise. Additionally, emerging approaches utilize behavioral analytics, contextual data, and continuous verification methods that monitor user activity and device behavior throughout a session, providing dynamic and adaptive authentication solutions. The goal of authentication protocols is to not only protect data confidentiality and system integrity but also to maintain a balance between security and user convenience across diverse platforms such as cloud services, IoT environments, and enterprise networks [20, 21].

Traditional cryptographic mechanisms, while effective in many contexts, often require high computation power and are not always convenient for limited devices such as drones [22]. Moreover, key storage and management remain persistent issues, especially in physically exposed systems. To address these challenges, improve security characteristics, and enable robustness against physical assaults, we provide a lightweight semi-centralized authentication scheme for drone-to-drone communication, leveraging the inherent security advantages of Physical Unclonable Functions (PUFs).

Generally, PUFs exploit the intrinsic manufacturing variations in hardware to generate unique and unpredictable responses to specific challenges, effectively serving as hardware fingerprints. This property makes PUFs highly attractive for secure device identification and authentication without storing sensitive keys in non-volatile memory [23].

In this paper, we present a novel mutual authentication protocol for UAVs that ensures secure and efficient verification between drones. Combining the Computational Diffie–Hellman problem with PUF, Our protocol is designed to resist common attack vectors while maintaining low computational and communication overhead.

As a result, it is particularly well-suited for real-time, resource-constrained aerial networks. To demonstrate the effectiveness of the proposed scheme in securing drone-to-drone communication, we evaluate it in terms of security robustness, and efficiency.

The rest of this chapter is organized as follows. Section two provides related works. Section three presents the system model and the proposed authentication scheme. Security analysis is provided in the fourth section. Section five examines the performance of our mechanism. Finally, section six concludes the chapter.

2 Related Works

Considering the importance of drone operations, securing communicated data in IoD networks has gained considerable attention in recent years. To ensure data integrity, confidentiality, and authenticity, different authentication protocols have been proposed [24–27]. Public key infrastructure (PKI)-based mechanisms are extensively exploited for mutual authentication; however, they usually require considerable computational power and energy, which makes them inadequate for lightweight drone environments. Symmetric key-based protocols, though more robust, often face obstacles in key distribution and storage, especially in dynamic networks [28].

In Venkatasamy Sureshkumar et al. [29], present a novel Medical Wireless Sensor network architecture and provide an authentication and key agreement scheme based on ECC. They prove the correctness of their scheme using Burrows–Abadi–Needham logic. Further analysis shows the robustness of the provided scheme.

In 2023, Mohammad Reza Servati et al. [30] conducted a cryptanalysis of Sureshkumar et al.'s protocol [29] and proved its vulnerability to traceability attacks. To overcome these vulnerabilities, they provided an enhanced authentication protocol called ECCbAS based on ECC and proved its robustness using the Scyther tool. Additionally, Yuanyuan Zhang et al. [31] proposed a lightweight batch authentication protocol for the IoD environments. The scheme's purpose is to permit user authentication with multiple drones in parallel. The security analysis of the provided protocol ensures its robustness against several attacks, including physical attacks, since it uses PUF.

In 2024, Anwar Ghani et al. [32] presented a lightweight authentication protocol built on the Computational Diffie–Hellman (CDH) for decentralized edge computing networks. They provided a peer-to-peer authentication protocol between two devices without using a third trust authority. Muhammed Tanveer et al. [33] provide a secure authentication scheme based only on PUF for IoD networks. The protocol proves its resiliency against impersonation, replay, and privileged insider assaults.

Earlier in 2025, Sunil Prajapat et al. [34] provided a new authentication protocol for Brain-Computer Interface systems that preserves privacy based on ECC. The proposed protocol is optimized for resource-constrained devices by balancing strong security with cost-effectiveness. Further, it shows a strong resilience against privileged insiders, password guessing, and user impersonation.

The analysis of recent authentication schemes shows that they have explored the integration of lightweight cryptographic mechanisms to reduce costs while maintaining acceptable levels of security. Elliptic Curve Cryptography (ECC) and hash-based mechanisms are prominent examples, yet they still depend on secure key management and may remain susceptible to physical attacks if keys are stored in insecure memory.

3 System Model

This section describes the architecture, assumptions, and threat model in our PUF-based semi-centralized drone-to-drone authentication protocol.

We consider a semi-centralized network of drones that contains a gateway used only for PUF challenge generation. Each drone is equipped with a unique embedded PUF module capable of generating response pairs based on a given challenge and intrinsic hardware characteristics.

Drones can initiate mutual authentication when they come into the communication range of one another. In every authentication request, the gateway computes the challenge using its secret key for each drone.

4 Provided Authentication Protocol

Our protocol consists of three stages. In this section, every stage is explained in detail. The channel used all along the registration stage is considered secure. However, during the authentication stage, the channel used is considered open. Table 1 illustrates the annotations utilized in the rest of the chapter.

4.1 Enrollment Stage

The gateway first chooses its GID, private key KG, and hash function h(.). After that, it released GID and h(.) and maintained KG in its storage.

4.2 Drone Registration Stage

Firstly, the drone chooses its identity IDi and its secret key xi and computes its public key *yi* as in Eq. (1), based on the CDH [24]. Then, it chooses a random nonce ra and calculates its pseudo-identity HID*i* as in Eq. (2). Lastly, the drone transmits the check-in request {HIDi, yi} to the GWN.

Table 1 Annotations

Notation	Description
ID_i, ID_j	Drones identities
HIDi, HIDj	Drones pseudo identities
GWN GID	Gateway Gateway identity
xi, xj	The drones' private keys
yi, yj	The drones public keys
KG	Gateway secret key
h(.)	One-way hashing function
SK	Session key
Gen, Rep	Generation and reproduction algorithm of fuzzy extractor
ri, rj	Random nonces
Ti	Timestamp
PUF	The Physical Unclonable Function
Ci, Bi	The challenge/ response pair
$\oplus$	Xor procedure
$\|\|$	String concatenation procedure

The GWN checks if the pseudo-identity is available when it receives the registration request. Then, it calculates the PUF challenge Ci as depicted in Eq. (3) and maintans {HIDi, yi} in its storage. Eventually, the gateway communicates {GID, Ci} to the drone and releases yi to all drones.

As fast as the drone obtains the feedback, it determines Bi, (Ri, Pi), Vi, and Xi as in Eq. (4)-(7). Then, it retains {HIDi, Xi, yi, Pi, Vi, GID} in its storage.

$$yi = g^{xi} \tag{1}$$

$$\text{HID}i = h(\text{ID}i||Ri) \tag{2}$$

$$Ci = h(\text{HID}i||KG) \tag{3}$$

$$Bi = \text{PUF}(Ci) \tag{4}$$

$$(Ri, Pi) = \text{Gen}(Bi) \tag{5}$$

$$Vi = h(Ci||Ri||\text{GID}) \tag{6}$$

$$Xi = xi \oplus h(\text{HID}i||Ri) \tag{7}$$

4.3 Authentication and Key Agreement Stage

The authentication stage requires drones to accomplish five strides, as stated in this section. Figure 1 illustrates the process of authentication and key agreement.

Step 1: The drone Di transmits the pseudo-identity HIDi and a timestamp T1 to the gateway to initiate the session. The gateway checks the validity of T1 upon receiving it from the drone Di and calculates the challenges Ci and Ai as per Eqs. (8) and (9). After that, it transmits {Ai, T2} to the drone Di.

$$Ci = h(\mathrm{HID}i||\mathrm{KG}) \tag{8}$$

$$Ai = Ci \oplus \mathrm{HID}i \oplus T2 \tag{9}$$

Step 2: Upon obtaining the message, the drone checks the T2's freshness, retrieves Ci* as in Eq. (10), and determines Bi, Ri*, and Vi* as in Eqs. (11)-(13). Subsequently, it checks if Vi* = Vi. After the verification of Vi*, the drone retrieves xi* as in Eq. (14), generates T3, chooses a random nonce ri, and calculates M1, M2, M3 as illustrated in Fig. 1. Eventually, the drone communicates {M1, M3, T3, HIDj} to the GWN.

$$Ci^* = A1 \oplus \mathrm{HID}i \oplus T2 \tag{10}$$

$$Bi = \mathrm{PUF}\left(Ci^*\right) \tag{11}$$

$$Ri* = \mathrm{Rep}(Bi, Pi) \tag{12}$$

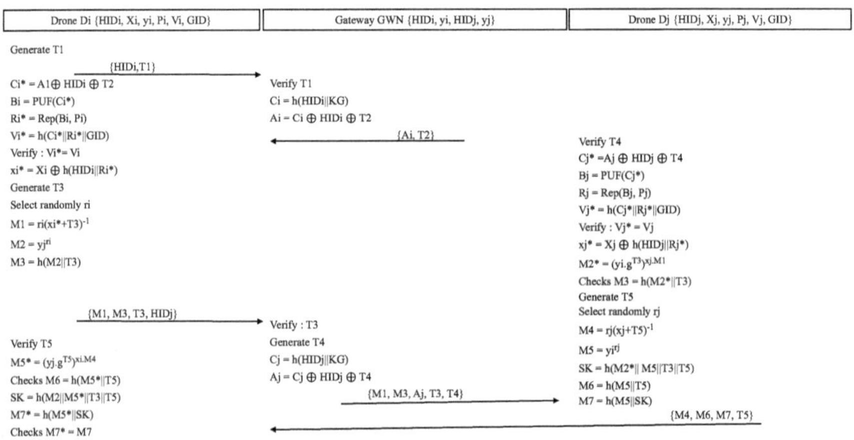

Fig. 1 Authentication and key agreement phase

$$Vi^* = h\left(Ci^*||Ri^*||GID\right) \tag{13}$$

$$xi^* = \mathrm{Xi} \oplus \mathrm{h(HIDi}||Ri^*) \tag{14}$$

Step 3: First, the GWN checks the validity of T3, generates T4, and determines the PUF challenge Cj and Aj as in Eq. (15) and (16). Subsequently, it transmits {M1, M3, Aj, T3, T4} to the drone Dj.

$$Cj = h(\mathrm{HID}j||KG) \tag{15}$$

$$Aj = Cj \oplus \mathrm{HID}j \oplus T4 \tag{16}$$

Step 4: After getting the authentication requisition, the drone Dj checks the validity of T4, retrieves Cj* as in Eq. (17), and calculates Bj, Rj, and Vj as in Eq. (18), (19) and (20). Eventually, it checks if Vj* = Vj. If it is true, the drone recuperates xj* and M2* as in Eq. (21) and (22) and verifies if Eq. (23) is true. If it is true, the drone generates T5, chooses a random nonce rj, and calculates M4, M5, M6, SK, and M7 as demonstrated in Fig. 1. Afterward, it transmits {M4, M6, M7, T5} to the drone Di.

$$Cj^* = Aj \oplus \mathrm{HID}j \oplus T4 \tag{17}$$

$$Bj = \mathrm{PUF}\left(Cj^*\right) \tag{18}$$

$$Rj = \mathrm{Rep}(Bj, Pj) \tag{19}$$

$$Vj^* = h\left(Cj^*||Rj^*||GID\right) \tag{20}$$

$$xj^* = Xj \oplus h(\mathrm{HID}j||Rj^*) \tag{21}$$

$$M2^* = \left(yi.g^{T3}\right)^{xj.M1} \tag{22}$$

$$M5 = h(M2^*||\mathrm{HID}i^*||ri||T3) \tag{23}$$

Step 5: When the drone receives the response message, it checks T5's novelty, calculates M5* as in Eq. (24), and verifies the propriety of Eq. (25). If it is true, the drone determines SK and M7* as in Eq. (26) and (27), and verifies if M7* = M7. If it is true, the authentication is established, and the secret key between the drones is the session key SK.

$$M5^* = \left(yi.g^{T5}\right)^{xi.M4} \tag{24}$$

$$\mathrm{M6} = \mathrm{h}(M5^*||\mathrm{T5}) \tag{25}$$

$$\mathrm{SK} = \mathrm{h}\left(\mathrm{M2}||M5^*||\mathrm{T3}||\mathrm{T5}\right) \tag{26}$$

$$M7^* = h(M5^*||\mathrm{SK}) \tag{27}$$

5 Security Analysis

To evaluate the security of our mechanism, we consider the Dolev-Yao threat with full control over the communication channel [35]. The attacker may eavesdrop, replay, forge, or modify messages and can attempt to impersonate legitimate drones.

Mutual authentication: Our system involves drones Di and Dj authenticating each other. The Dj checks that Di is legitimate by analyzing the recovered message M3* with the M3 transmitted by the drone Di. To recover M3*, the drone Dj utilizes his secret key xj as depicted in Fig. 1. As a result, no third entity can attest the drone Di. Similarly, to validate the drone Dj, the drone Di checks the validity of the recovered M6* and M7* using M6 and M7 obtained from Dj.

Anonymity and untraceability: Our protocol exchanges hashed identities forms. Given that the opponent cannot get private keys, he may not estimate M2 or M5. Consequently, he is unable to recover HIDi or HIDj. Even if the attacker has benefit access to HIDi or HIDj, extracting IDi or IDj will be impossible since the original input cannot be retrieved from the hash output. Additionally, the drone computes an authentication demand that varies from session to session because of random numbers.

Perfect secrecy: Considering the session key in our protocol is computes as SK = h(M2||M5||T3||T5), the opponent requires ri and rj to estimate it. Nevertheless, the CDH irreversibility prevents the opponent from retrieving ri and rj from the communicated data. Moreover, ri and rj are unpredictable nones selected arbitrarily in every session. As a result, the opponent may not alter earlier or future keys even when he gain access to a session key.

Impersonation attack: To impersonate the drone Di, the attacker should construct a valid request {M1, M3, M4, M5, T3, yj}. Assuming the attacker sends HIDi to the gateway and recovers Ai, the attacker may not reproduce Ri or recover the private key xi to construct a valid pair (M2, M5). The same things happen if the opponent tries to impersonate the drone Dj. For this reason, our scheme is safe against impersonation attacks.

Replay attack: The first step in each authentication step is to check the timestamp's novelty. If it is not fresh, the entity declines the authentication request. On the other hand, when the opponent attempts to use fresh timestamps, he needs to estimate M1 and M5, which is unfeasible since he does not have secret arguments.

DDoS attack: Since the messages exchanged in our protocol contain fresh timestamps, the authentication may be declined as fast as the entity verifies the timestamp's validity. The opponent may not use previous authentication requests nor estimate valid ones. As a result, our protocol is robust against DDoS.

Denning-Sacco attack: In our mechanism the session kay is computed as SK = h(M2||M5||T3||T5), where M2, M5, T3,T5 differ by the session change. İt does not use any long-term argument, which means it is resistant to Denning-Sacco attack.

Stolen verifier: Since the gateway verification table contains only identities and public keys, even if the attacker accesses it, he may not estimate the session key nor retrieve secret keys xi and xj.

Man-in-the-middle attack: Considering that the attacker cannot recover ri and xi, he may not falsify an intercepted authentication requisition. For this reason, our mechanism is unsusceptible to man-in-the-middle assault.

Physical attack: Generally, the physical attack may lead the attacker to access the data preserved in the drone's storage. Nevertheless, the drone PUF is not imitable. Therefore, the tampered drone cannot be authenticated, given that he may not estimate the response Ri = PUF(Ci).

6 Performance Evaluation

This section conducts a performance analysis of the provided authentication protocol and evaluates it against state-of-the-art protocols.

6.1 Security Achievements

Table 2 illustrates the result of this comparative analysis, highlighting the varying levels of protection and feature support among existing solutions. Protocols [29–31], and [34] demonstrate strong foundational security by supporting all six core features, including mutual authentication, anonymity, unlinkability, key agreement, key secrecy, and perfect forward secrecy. However, their resistance to attacks is limited. The protocols [29] and [34] resist only impersonation and replay assaults, while the protocol [30] provides imperfect robustness, missing resistance upon physical attacks, stolen verifier attacks, and Denning-Sacco vulnerabilities. The protocol

Table 2 Security performance

Protocol	F_1	F_2	F_3	F_4	F_5	F_6	A_1	A_2	A_3	A_4	A_5	A_6	A_7
[29]	✓	✓	✓	✓	✓	✓	✓	✓	–	–	–	–	–
[30]	✓	✓	✓	✓	✓	✓	✓	✓	–	✓	–	–	✓
[31]	✓	✓	✓	✓	✓	✓	✓	–	✓	–	–	–	✓
[32]	✓	✗	✗	✓	–	–	✓	✓	–	✓	✓	–	✓
[33]	✓	✓	✓	✓	✓	–	✓	✓	✓	✓	–	–	✓
[34]	✓	✓	✓	✓	✓	✓	✓	✓	–	–	–	–	–
Our protocol	✓	✓	✓	✓	✓	✓	✓	✓	✓	✓	✓	✓	✓

F_1: mutual authentication, F_2: Anonymity, F_3: unlinkability, F_4: key agreement, F_5: key secrecy, F_6: perfect forward secrecy, A_1: Impersonation attack, A_2: reply attack, A_3: physical attack, A_4: DoS attack, A_5: Stolen verifier, A_6: Denning-sacco attack, A_7: man-in-the-middle.

[31] does not consider replay attacks, DoS attacks, stolen verifiers, and Denning-sacco attacks. Protocol [32] shows the weakest feature set, lacking anonymity, unlinkability, and key secrecy. However, it resists several assaults, including impersonation, replay, DoS, stolen verifier, and man-in-the-middle attacks. Protocol [33] offers a balanced performance but lacks perfect forward secrecy and remains vulnerable to kind of cryptographic threats, including stolen verifiers and Denning-sacco attacks. In contrast, the proposed protocol outperforms all the state-of-the-art ones by fully supporting every essential security feature and resisting all seven major attack vectors, making it the most robust and comprehensive solution among the compared approaches.

6.2 Computation Cost

Table 4 demonstrates the computation costs of the authentication protocols compared. Table 3 illustrates the annotation and execution time requirements of the operations utilized in the proposed mechanism based on [32] and [36]. We ignored the XOR and concatenation operations regarding their neglected execution time.

The computation cost comparison reveals significant differences in the efficiency of the analyzed protocols. Protocols [29, 30], and [34] show relatively low computation costs of 15.188 ms, 15.63 ms, and 15.671 ms, respectively. However, they

Table 3 Cryptographic run time

Operation	Run time (ms)
Power exponentiation T_E	1.864
One-way hashing T_H	0.0025
Fuzzy extractor T_{FE}	0.0098
PUF operation T_{PUF}	0.0021

Table 4 Computation cost comparison

Protocol	Computation cost (ms)
[29]	15.188
[30]	15.63
[31]	38.01
[32]	42.386
[33]	21.638
[34]	15.671
Our protocol	7, 5

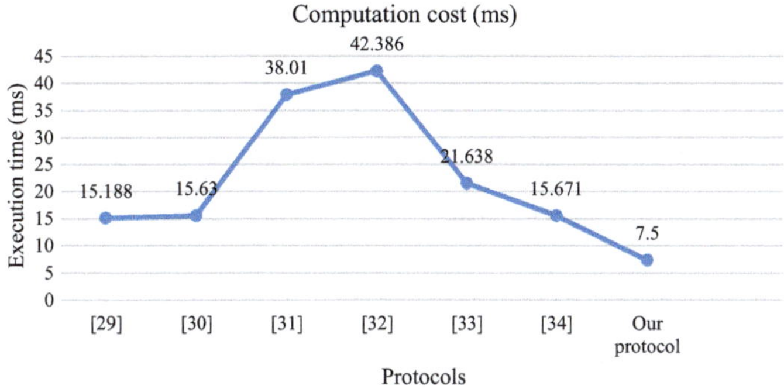

Fig. 2 Protocols computation overheads

ensure a limited security level. Protocols [31, 32], and [33] require much higher computation costs, 38.01 ms, 42.386 ms, and 21.638 ms, respectively. While they are vulnerable to various attacks, these computation overheads make them unsuitable for resource-constrained environments and real-time use cases. Our provided scheme requires only 7.5 ms, which makes it the most efficient of all the solutions compared. Figure 2 illustrates the run time requirements for each authentication scheme.

6.3 Communication Cost

Table 5 shows the communication costs of the authentication protocols compared. Table 5 illustrates the annotation and size of the parameters utilized in the proposed mechanism based on [32] and [36].

As pictured in Table 6, the protocol [29] requires the highest communication cost of 4416 bits, followed by the protocol [30] with 3456 bits. Protocols [32] and [33] show moderate communication costs of 2112 and 2272 bits, respectively, offering better efficiency but still not optimal. Protocol [31] requires the lowest communication cost of just 832 bits, making it highly lightweight in terms of data transmission.

Table 5 Parametric sizes

Parameter	Size (bit)
Identity	32
Random number	128
Timestamp	32
Hash function	160

Table 6 Communication cost comparison

Protocol	Communication cost (bits)
[29]	4416
[30]	3456
[31]	832
[32]	2112
[33]	2272
[34]	1632
Our protocol	1856

However, it is limited in terms of attack resistance and practical security. Similarly, the protocol [34] achieves a relatively low cost of 1632 bits while it presents vulnerabilities to various attacks. In contrast, the proposed protocol maintains a well-balanced communication cost of 1856 bits, significantly lower than the more complex protocols [29] and [30], while offering far stronger security. Figure 3 illustrates the communication requirements for each authentication scheme.

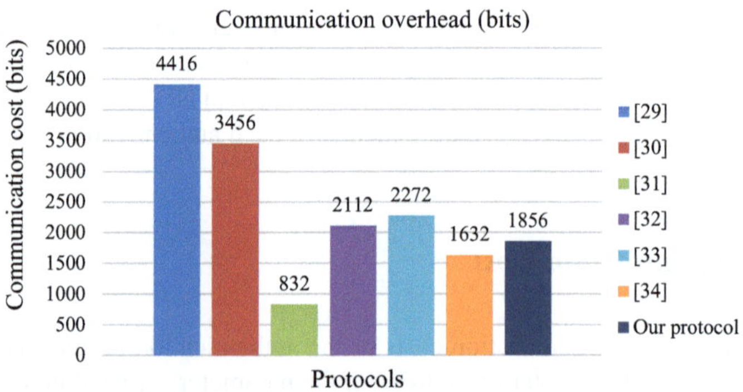

Fig. 3 Protocols communication overheads

7 Conclusion

Frequently, IoD networks are susceptible to multiple assaults because of the outdoor use of drones. This paper aims to create an authentication system that uses the CDH problem and PUF to secure drone-to-drone communications in such networks. The proposed protocol's security assessment demonstrates its resiliency to common raids. Moreover, our mechanism evaluation regarding the state-of-the-art ones shows that the proposed solution effectively minimizes computation and communication overheads. This combination of high security and low costs makes it exceptionally well-suited for practical, real-world IoD deployment. However, it is possible that the protocol can encounter difficulties in managing large networks with thousands of drones. Furthermore, the drone's energy consumption might be affected by the computational cost, which could shorten their functional lifetime. Subsequent research should aim to enhance the authentication mechanism by minimizing energy consumption while preserving both security and scalability.

References

1. Shah, I.A., Jhanjhi, N.Z., Ujjan, R.M.A.: Drone Technology in the Context of the Internet of Things. In: Cybersecurity Issues and Challenges in the Drone Industry. pp. 88–107. IGI Global (2024).
2. Mandloi, D., Arya, R., Verma, A.K.: Internet of Drones. In: Arya, R., Sharma, S.C., Verma, A.K., and Iyer, B. (eds.) Recent Trends in Artificial Intelligence Towards a Smart World. pp. 353–373. Springer Nature Singapore, Singapore (2024). https://doi.org/10.1007/978-981-97-6790-8_13.
3. Bine, L.M., Boukerche, A., Ruiz, L.B., Loureiro, A.A.: Connecting internet of drones and urban computing: Methods, protocols and applications. Computer Networks. 239, 110136 (2024).
4. Azrour, M., Dargaoui, S., Mabrouki, J., Guezzaz, A., Benkirane, S., Shafik, W., Ahmad, S.: A Survey of Machine and Deep Learning Applications in the Assessment of Water Quality. In: Mabrouki, J. and Mourade, A. (eds.) Technical and Technological Solutions Towards a Sustainable Society and Circular Economy. pp. 471–483. Springer Nature Switzerland, Cham (2024). https://doi.org/10.1007/978-3-031-56292-1_38.
5. Dargaoui, S., Azrour, M., El Allaoui, A., Guezzaz, A., Benkirane, S., Alabdulatif, A., Amounas, F.: Applications of Blockchain in Healthcare:: Review Study. IoT, Machine Learning and Data Analytics for Smart Healthcare. 1–12.
6. Mohy-eddine, M., Guezzaz, A., Benkirane, S., Azrour, M.: Malicious detection model with artificial neural network in IoT-based smart farming security. Cluster Computing. 27, 7307–7322 (2024).
7. Blali, A., Dargaoui, S., Azrour, M., Guezzaz, A., Amounas, F., Alabdulatif, A.: Analysis of deep learning-based intrusion detection systems in IoT environments. EDPACS. 1–35 (2025).
8. Hazman, C., Guezzaz, A., Benkirane, S., Azrour, M.: A smart model integrating LSTM and XGBoost for improving IoT-enabled smart cities security. Cluster Computing. 28, 70 (2025).
9. Mohy-Eddine, M., Guezzaz, A., Benkirane, S., Azrour, M., Bella, K.: A monitoring system with deep learning for IoT smart environments security. In: Recent Advances in Internet of Things Security. pp. 93–104. CRC Press (2025).
10. Assmi, H., Guezzaz, A., Benkirane, S., Azrour, M., Jabbour, S., Innab, N., Alabdulatif, A.: A Robust Security Detection Strategy for Next Generation IoT Networks. Computers, Materials & Continua. 82, (2025).

11. Dargaoui, S., Azrour, M., Mabrouki, J., El allaoui, A., Guezzaz, A., Benkirane, S., Alabdulatif, A.: Security Issues in Internet of Medical Things. Presented at the December 19 (2023). https://doi.org/10.1201/9781003438779-5.
12. Dargaoui, S., Azrour, M., El Allaoui, A., Guezzaz, A., Alabdulatif, A., Alnajim, A.: An Exhaustive Survey on Authentication Classes in the IoT Environments. Indonesian Journal of Electrical Engineering and Informatics (IJEEI). 12, (2024).
13. Dargaoui, S., Azrour, M., El Allaoui, A., Guezzaz, A., Benkirane, S., Alabdulatif, A., Amounas, F.: IoT-Driven Smart Agriculture: Security Issues and Authentication Schemes Classification. In: Mejdoub, Y. and Elamri, A. (eds.) Proceeding of the International Conference on Connected Objects and Artificial Intelligence (COCIA2024). pp. 61–66. Springer Nature Switzerland, Cham (2024). https://doi.org/10.1007/978-3-031-70411-6_10.
14. Dargaoui, S., Azrour, M., El Allaoui, A., Guezzaz, A., Benkirane, S., Alabdulatif, A., Amounas, F.: Internet-of-Things-Enabled Smart Agriculture: Security Enhancement Approaches. In: 2024 4th International Conference on Innovative Research in Applied Science, Engineering and Technology (IRASET). pp. 1–5 (2024). https://doi.org/10.1109/IRASET60544.2024.10548705.
15. Dargaoui, S., Azrour, M., El allaoui, A., Guezzaz, A., Alabdulatif, A., Alnajim, A.: Internet of Things Authentication Protocols: Comparative Study. Computers, Materials & Continua. 1–10 (2024). https://doi.org/10.32604/cmc.2024.047625.
16. Dargaoui, S., Azrour, M., El Allaoui, A., Guezzaz, A., Benkirane, S.: Authentication in Internet of Things: State of Art. In: Proceedings of the 6th International Conference on Networking, Intelligent Systems & Security. pp. 1–6 (2023).
17. Abu-Zaid, A., Aljaidi, M., Al-Na'amneh, Q., Samara, G., Alsarhan, A., Qadoumi, B.: Advancements and Challenges in the Internet of Drones Security Issues: A Comprehensive Review. Machine Intelligence Applications in Cyber-Risk Management. 1–24 (2025).
18. Shah, I.A., Jhanjhi, N.Z., Ray, S.K.: IoT Devices in Drones: Security Issues and Future Challenges. In: Cybersecurity Issues and Challenges in the Drone Industry. pp. 217–235. IGI Global (2024).
19. Dargaoui, S., Azrour, M., Allaoui, A.E., Guezzaz, A., Benkirane, S.: Machine learning for security boosting in Internet of Things environments. In: Recent Advances in Internet of Things Security. CRC Press (2025).
20. Kokila, M., Reddy, S.: Authentication, access control and scalability models in Internet of Things Security–A review. Cyber Security and Applications. 3, 100057 (2025).
21. Jeelani, M., Muqeem, M., Ahmad, S., Zafar, A., Azrour, M., Singh, K.P.: Wireless Sensor Networks and Communication Protocols. In: Integrating Intelligent Control Systems With Sensor Technologies. pp. 189–238. IGI Global Scientific Publishing (2025).
22. Dargaoui, S., Azrour, M., El Allaoui, A., Amounas, F., Guezzaz, A., Attou, H., Hazman, C., Benkirane, S., Bouazza, S.H.: An Overview of the Security Challenges in IoT Environment. In: Mabrouki, J., Mourade, A., Irshad, A., and Chaudhry, S.A. (eds.) Advanced Technology for Smart Environment and Energy. pp. 151–160. Springer International Publishing, Cham (2023). https://doi.org/10.1007/978-3-031-25662-2_13.
23. Dargaoui, S., Azrour, M., Allaoui, A.E., Guezzaz, A., Hammoudeh, M.A.A.: New Efficient PUF-Based Authentication Protocol for IoT-Driven Smart Agriculture. JAIT. 16, 582–593 (2025). https://doi.org/10.12720/jait.16.4.582-593.
24. Dargaoui, S., Azrour, M., El Allaoui, A., Guezzaz, A., Alabdulatif, A., Ahmad, S.: ECC-Based Anonymous and Multi-factor Authentication Scheme for IoT Environment. International Journal of Online & Biomedical Engineering. 21, (2025).
25. Shariq, M., Conti, M., Singh, K., Dwivedi, S.K., Abdussami, M., Amin, R., Masud, M.: Design of Provably Secure and Lightweight Authentication Protocol for Unmanned Aerial Vehicle systems. Computer Communications. 228, 107971 (2024).
26. Zhang, Z., Hsu, C., Au, M.H., Harn, L., Cui, J., Xia, Z., Zhao, Z.: Prlap-iod: A puf-based robust and lightweight authentication protocol for internet of drones. Computer networks. 238, 110118 (2024).

27. Anwar, T., Khan, G.A., Ashraf, Z., Ansari, Z.A., Ahmed, R., Azrour, M.: The combination of blockchain and the internet of things (iot): applications, opportunities, and challenges for industry. Blockchain and Machine Learning for IoT Security. 56–76 (2024).
28. Attou, H., Guezzaz, A., Benkirane, S., Azrour, M.: A New Secure Model for Cloud Environments Using RBFNN and AdaBoost. SN COMPUT. SCI. 6, 188 (2025). https://doi.org/10.1007/s42979-025-03691-1.
29. Sureshkumar, V., Amin, R., Vijaykumar, V.R., Sekar, S.R.: Robust secure communication protocol for smart healthcare system with FPGA implementation. Future Generation Computer Systems. 100, 938–951 (2019).
30. Servati, M.R., Safkhani, M.: ECCbAS: An ECC based authentication scheme for healthcare IoT systems. Pervasive and Mobile Computing. 90, 101753 (2023).
31. Zhang, Y., Meng, L., Zhang, M., Meng, W.: A secure and lightweight batch authentication scheme for Internet of Drones environment. Vehicular Communications. 44, 100680 (2023).
32. Ghani, A., Jan, S.U., Chaudhry, S.A., Ahmad, R., Kim, D.-H.: MCDH-SLKAP: Modified Computational Diffie-Hellman based Secure and Lightweight Key Agreement Protocol for Decentralized Edge Computing Networks. IEEE Access. (2024).
33. Tanveer, M., Aldosary, A., Kumar, N., Aldossari, S.A.: SEAF-IoD: Secure and efficient user authentication framework for the Internet of Drones. Computer Networks. 247, 110449 (2024).
34. Prajapat, S., Rana, A., Kumar, P., Das, A.K., Susilo, W.: Privacy-preserving authentication protocol for user personal device security in Brain–Computer Interface. Computer Standards & Interfaces. 94, 104009 (2025).
35. Dolev, D., Yao, A.: On the security of public key protocols. IEEE Transactions on information theory. 29, 198–208 (1983).
36. Nyangaresi, V.O., Al-Joboury, I.M., Al-sharhanee, K.A., Najim, A.H., Abbas, A.H., Hariz, H.M.: A biometric and physically unclonable function–Based authentication protocol for payload exchanges in internet of drones. e-Prime-Advances in Electrical Engineering, Electronics and Energy. 7, 100471 (2024).

Detecting Illicit Bitcoin Transactions with Graph Neural Networks and Adaptive Filtering

Charaf Hamidi, Mohamed Badiy, Hind Aiouej, karim El Moustafid, Salma Gaou, Fatima Amounas, and Hicham Tribak

Abstract The increasing complexity of financial transactions driven by global economic integration and advancements in technology has made fraudulent activities increasingly challenging to detect. This paper presents an innovative approach leveraging Graph Convolutional Networks (GCNs) to detect financial fraud within the Elliptic Bitcoin Dataset. By combining distributed data preprocessing, advanced graph construction, and feature engineering techniques, our method captures complex relational patterns and temporal dynamics inherent in transaction networks. The proposed solution leverages GCNs to enhance the detection of fraudulent transactions while addressing challenges such as data imbalance and feature weakening through adaptive filtering mechanisms. Experimental validation demonstrates the robustness of the proposed model, achieving an accuracy of 95% on the dataset, while highlighting challenges related to minority class detection. This paper discusses the implications of these findings and proposes future research directions to improve fraud detection systems.

C. Hamidi (✉) · H. Aiouej · S. Gaou
Laboratory of Engineering Sciences, Faculty of Science, University Ibn Zohr, Agadir, Agadir, Morocco
e-mail: charaf.hamidi.24@edu.uiz.ac.ma

M. Badiy · F. Amounas
IMIA Laboratory, MSIA Team, Faculty of Sciences and Techniques Errachidia, Moulay Ismail University of Meknes, Meknes, Morocco
e-mail: m.badiy@edu.umi.ac.ma

F. Amounas
e-mail: f.amounas@umi.ac.ma

E. Moustafid
Laboratory for Legal Research and Policy Analysis (LAREJAP), University Cadi Ayyad, Marrakech, Morocco
e-mail: K.elmoustafid@uca.ac.ma

H. Tribak
Physics, Energy, and Information Processing, Multidisciplinary Faculty, University Ibn Zohr, Ouarzazate, Morocco
e-mail: h.tribak@uiz.ac.ma

M. Azrour and A. Alabdulatif (eds.), *AI-Driven Security for Next-Generation IoT Systems*, https://doi.org/10.1007/978-3-032-08784-3_11

Keywords Financial fraud detection · Graph neural networks · Graph convolutional networks · Cryptocurrency fraud

1 Introduction

Cryptocurrencies have significantly reshaped global financial transactions by providing decentralized, transparent, and secure alternatives to traditional currency systems. Despite these advantages, the pseudonymous nature of blockchain technology has also facilitated an increase in illicit activities, such as financial fraud and money laundering [1–3]. Detecting fraudulent cryptocurrency transactions has therefore become an urgent and challenging task for researchers and practitioners alike. Traditional fraud detection approaches primarily rely on transaction-centric analyses that neglect relational and structural information inherent in transaction networks [4]. Recently, Graph Neural Networks (GNNs), especially Graph Convolutional Networks (GCNs), have emerged as promising solutions, capable of capturing complex relational patterns and interactions within graph-structured data [5–7].

The primary objective of this study is to evaluate the efficacy of GCNs in identifying fraudulent transactions within blockchain networks. To achieve this, we utilize the widely recognized Elliptic Bitcoin Dataset, characterized by detailed transactional attributes and labels explicitly identifying illicit and licit activities. We aim to demonstrate that graph-based techniques can significantly outperform traditional models by effectively leveraging structural, temporal, and contextual transaction information. Our contributions are as follows: (1) detailed preprocessing and advanced graph construction methods tailored specifically for cryptocurrency data, (2) a robust evaluation of GCNs addressing critical challenges such as data imbalance and node feature relevance, and (3) identification of actionable insights for improving fraud detection performance.

2 Related Work

Financial fraud detection, particularly within cryptocurrency networks, has garnered considerable research interest due to the escalating complexity and volume of fraudulent activities [8–10]. Early methodologies heavily relied on traditional supervised learning algorithms, including logistic regression, support vector machines, and decision trees [11–13]. However, these methods are limited in their capacity to utilize complex relational data, as they primarily depend on transaction-level attributes and fail to incorporate the rich structural context of transactional networks.

In recent years, the application of graph-based methods, specifically Graph Neural Networks (GNNs), has significantly advanced the detection capabilities in this domain. GNNs effectively utilize structural relationships inherent in data represented as graphs, making them highly suitable for analyzing complex financial transactions

[14–16]. Among various GNN architectures, Graph Convolutional Networks (GCNs) have been particularly effective due to their simplicity and powerful representation learning capabilities [17–20].

Several studies have demonstrated the potential of GCNs in identifying illicit transactions within blockchain networks. For instance, Weber et al. [21] utilized graph convolutional approaches to analyze Bitcoin transaction data, achieving notable improvements in fraud detection accuracy compared to classical methods. Similarly, Liu et al. [6] explored enhanced graph embedding techniques, integrating temporal dynamics to better capture evolving fraudulent patterns. Furthermore, recent works have tackled critical challenges such as class imbalance and scalability, employing techniques like adaptive sampling, class-weighted loss functions, and distributed graph processing frameworks [1, 5].

In summary, existing research underscores the efficacy and potential of graph neural network approaches, particularly GCNs, for financial fraud detection tasks. However, challenges remain regarding data imbalance, scalability, and effective incorporation of temporal dynamics. Addressing these issues forms the basis for our approach in this study.

3 Methodology

We utilized the Elliptic cryptocurrency transaction dataset to detect fraudulent activities. This dataset includes transaction-level features comprising 93 transaction-based and 72 aggregated features, labeled explicitly as either illicit or licit, along with an edge list representing transaction flows. Initial preprocessing involved merging transaction features with class labels, filtering out transactions labeled as “unknown,” and eliminating irrelevant or duplicated entries. The processed data was then structured into a graph format, employing transaction IDs as nodes and their corresponding transaction flows as edges.

For our classification task, we employed a Graph Convolutional Network (GCN) architecture with two convolutional layers. The first layer comprised 16 hidden channels activated through ReLU functions, while the second layer directly output class logits. We trained the model using the Adam optimizer with a learning rate of 0.01 and incorporated L2 regularization (weight decay set at 5e-4). Training was conducted over 100 epochs, optimizing against the cross-entropy loss.

The dataset was partitioned into training, validation, and testing subsets, comprising 80%, 10%, and 10% of the data, respectively. The evaluation of the model's performance was measured using accuracy, precision, recall, and F1-score metrics.

4 Results

The performance of the GCN model was rigorously analyzed over the course of training. The training loss exhibited a steady decrease from 0.2264 at epoch 10 to 0.1253 at epoch 100, indicating effective model convergence. Correspondingly, training accuracy improved consistently, achieving 96.44% by the final epoch. Validation accuracy showed a parallel increase, reaching a peak of 96.41% at epoch 90 and slightly dropping to 96.39% at epoch 100, suggesting minimal overfitting (Fig. 1).

Table 1 presents a detailed classification report of the test dataset, highlighting a high overall accuracy of 95%. Precision, recall, and F1-score metrics revealed the model's superior performance, particularly for licit transactions, with precision at 96%, recall at 97%, and an F1-score of 96%. Although the model performed commendably on illicit transactions as well, achieving a precision of 92% and recall of 88%, this slightly lower recall indicates some difficulty in capturing all fraudulent cases, potentially due to inherent class imbalance.

The confusion matrix (Fig. 2) further illustrates the model's capability, clearly displaying the distribution of true positives and negatives, and a modest number of false negatives and false positives. This reinforces the earlier observed strengths and highlights areas for targeted improvements, such as addressing class imbalance to enhance detection sensitivity for illicit transactions. Overall, these results underline the suitability of GCNs in effectively classifying cryptocurrency transactions and suggest potential for further refinement through techniques addressing data imbalance and exploring deeper or alternative GNN architectures.

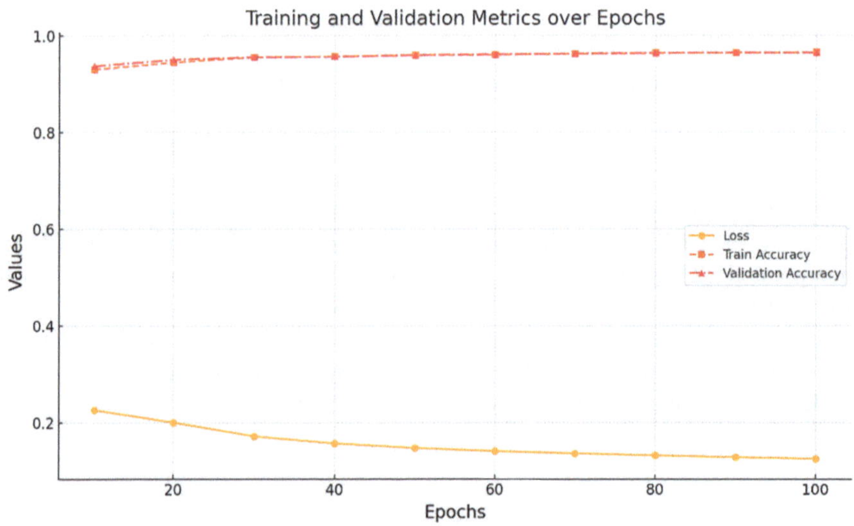

Fig. 1 Simulation training loss, training accuracy, and validation accuracy over epochs

Table 1 Classification report for GCN model

Class	Precision	Recall	F1-score	Support
Illicit	0.92	0.88	0.90	1628
Licit	0.96	0.97	0.96	8143
Accuracy			0.95	9771
Macro Avg	0.94	0.93	0.93	9771
Weighted Avg	0.95	0.95	0.95	9771

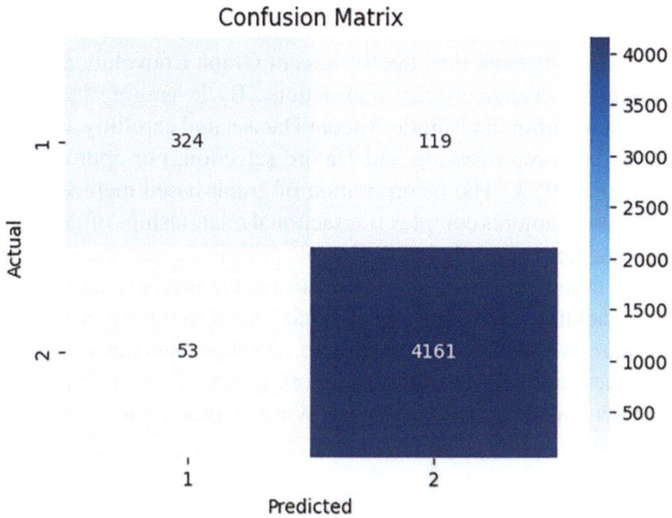

Fig. 2 Confusion matrix illustrating GCN classification outcomes

5 Discussion

The findings from our analysis highlight the considerable efficacy of Graph Convolutional Networks in detecting fraudulent cryptocurrency transactions. The consistently declining training loss alongside rising accuracy metrics indicates that our model successfully learned meaningful patterns within the transaction data. Notably, the minimal gap between validation and training accuracies demonstrates robust generalization capability and resilience against overfitting.

Despite high overall performance, the model exhibited a slightly lower recall for illicit transactions, indicating occasional failures in identifying some fraudulent transactions. This challenge is likely influenced by the inherent class imbalance within the dataset, where licit transactions significantly outnumber illicit ones. Future studies should explore data augmentation methods, class weighting strategies, or specialized loss functions to better address this imbalance.

Furthermore, the confusion matrix analysis provided clear insights into the distribution of correct and incorrect predictions, highlighting opportunities to enhance sensitivity toward fraudulent transactions. Considering the complex and evolving nature of cryptocurrency fraud, it would also be beneficial to investigate deeper neural network architectures or alternative graph-based methodologies that could capture more intricate transactional patterns.

6 Conclusion

This study has demonstrated the effectiveness of Graph Convolutional Networks in detecting fraudulent cryptocurrency transactions. By leveraging the inherent relational information within the Elliptic Bitcoin Dataset and carefully addressing challenges such as data preprocessing and feature selection, our approach achieved a notable accuracy of 95%. The incorporation of graph-based methods proves especially valuable, as it captures complex transactional relationships often overlooked by classical machine learning techniques. Despite promising outcomes, the slight drop in recall for the minority (illicit) class underscores the need for continued research, particularly in tackling class imbalance via advanced sampling or loss-adjustment strategies. Future work may explore deeper GNN architectures, temporal graph models, and other sophisticated neural network structures to further improve fraud detection and generalizability in rapidly evolving cryptocurrency environments.

References

1. Liu, X.F., Jiang, X.J., Liu, S.H., et al.: Knowledge discovery in cryptocurrency transactions: A survey. IEEE Access 9, 37229–37254 (2021).
2. Krishnan, L.P., Vakilinia, I., Reddivari, S., Ahuja, S.: Scams and solutions in cryptocurrencies—A survey analyzing existing machine learning models. Information 14(3), 171 (2023).
3. Pérez-Cano, V., Jurado, F.: Fraud detection in cryptocurrency networks—An exploration using anomaly detection and heterogeneous graph transformers. Future Internet 17(1), 44 (2025).
4. Abdallah, A., Maarof, M.A., Zainal, A.: Fraud detection system: A survey. J. Netw. Comput. Appl. 68, 90–113 (2016).
5. Wu, Z., Pan, S., Chen, F., Long, G., Zhang, C., Yu, P.S.: A comprehensive survey on graph neural networks. IEEE Trans. Neural Netw. Learn. Syst. 32(1), 4–24 (2021).
6. Liu, J., Li, W., Yu, P.S.: A survey of graph neural networks for anomaly detection in financial transactions. ACM Comput. Surv. (CSUR) 54(6), 1–27 (2022).
7. Zhong, M., Lin, M., Zhang, C., et al.: A survey on graph neural networks for intrusion detection systems: methods, trends and challenges. Comput. Secur. 2024, 103821.
8. West, J., Bhattacharya, M.: Intelligent financial fraud detection: a comprehensive review. Comput. Secur. 57, 47–66 (2016).
9. Al-Hashedi, K.G., Magalingam, P.: Financial fraud detection applying data mining techniques: A comprehensive review from 2009 to 2019. Comput. Sci. Rev. 40, 100402 (2021).
10. Motie, S., Raahemi, B.: Financial fraud detection using graph neural networks: A systematic review. Expert Syst. Appl. 240, 122156 (2024).

11. Alarab, I., Prakoonwit, S., Nacer, M.I.: Comparative analysis of supervised learningmethods for financial fraud detection. Expert Syst. Appl. 161, 113710 (2020).
12. Lorenz, J., Silva, J., De Faria, E.R.: Machine learning techniques for credit card fraud detection: A comparative analysis. J. Big Data 7(1), 1–21 (2020).
13. Lee, C.-W., Fu, M.-W., Wang, C.-C., Azis, M.I.: Evaluating machine learning algorithms for financial fraud detection: insights from Indonesia. Mathematics 13(4), 600 (2025).
14. Velickovic, P., Cucurull, G., Casanova, A., Romero, A., Liò, P., Bengio, Y.: Graph attention networks. Proc. Int. Conf. Learn. Represent. (2018).
15. Hiremath, A.C., Arya, A., Sriranga, L., Reddy, K.V.S.R., Nikhil, M.: Ensemble of graph neural networks for enhanced financial fraud detection. Proc. IEEE Int. Conf. Converg. Technol. (I2CT), 1–8 (2024).
16. Garg, A., Agarwal, A., Kaul, G., Singhal, A.: Leveraging graph neural networks for improved fraud detection in financial transactions. Proc. Int. Conf. Comput. Intell. Commun. Technol. Netw. (CICTN), 175–180 (2025).
17. Kipf, T.N., Welling, M.: Semi-supervised classification with graph convolutional networks. Proc. Int. Conf. Learn. Represent. (2017).
18. Liu, G., Tang, J., Tian, Y., Wang, J.: Graph neural network for credit card fraud detection. Proc. Int. Conf. Cyber-Phys. Soc. Intell. (ICCSI), 1–6 (2021).
19. Cheng, D., Zou, Y., Xiang, S., et al.: Graph neural networks for financial fraud detection: a review. Front. Comput. Sci. 19, 199609 (2025).
20. Kadam, P.: Financial fraud detection using jump-attentive graph neural networks. arXiv preprint arXiv:2411.05857 (2024).
21. Weber, M., Domeniconi, G., Chen, J., Weidele, D.K., Bellei, C., Robinson, T., Leiserson, C.E.: Anti-money laundering in Bitcoin: Experimenting with graph convolutional networks for financial forensics. KDD Workshop Anomaly Detect. Finance (2019).

Using Blockchain Technology to Secure Health Data in Telehealth

Salma Sabrou, Azidine Guezzaz, Said Jabbour, Said Benkirane, and Mourade Azrour

Abstract With the rise of artificial intelligence in the healthcare domain, telehealth is proving to be an invaluable tool for diagnosis, treatment, follow-up care, and health monitoring. Telehealth provides a simple and accessible solution for patients, particularly those living in rural or underserved regions, to consult with healthcare professionals and receive care. However, the exchange of sensitive health information in these digital environments raises serious concerns about data privacy, confidentiality, and integrity. This research investigates the application of blockchain technology to secure health data within telehealth systems, highlighting its key features and advantages. We propose a blockchain-based ecosystem designed to ensure data confidentiality, integrity, and controlled access.

Keywords Telehealth · Blockchain · Data security · Health data

---

S. Sabrou (✉) · A. Guezzaz · S. Benkirane
LaRTiD Laboratory, SISAR Team, Higher School of Technology, Cadi Ayyad University, Marrakesh, Morocco
e-mail: ssalma.sabrou@gmail.com

A. Guezzaz
e-mail: a.guzzaz@gmail.com

S. Benkirane
e-mail: sabenkirane@gmail.com

S. Jabbour
CRIL-CNRS, Artois University, Lens, France
e-mail: jabbour@cril.fr

M. Azrour
IMIA Laboratory, Faculty of Sciences and Technics, MSIA Team, Moulay Ismail University, Meknes, Morocco
e-mail: azrour.mourade@gmail.com

M. Azrour and A. Alabdulatif (eds.), *AI-Driven Security for Next-Generation IoT Systems*, https://doi.org/10.1007/978-3-032-08784-3_12

1 Introduction

Health systems are very important for people's health, the economy, and fairness in society. A good health system gives people medical care and also helps prevent illness, teaches people about health, and reduces differences in health. The World Health Organization states that a robust health system requires sufficient financial support, trained healthcare professionals, availability of essential drugs and tools, and reliable health data [1]. Research suggests that poor healthcare results in more deaths than having no healthcare at all, especially in poorer countries. Improving healthcare quality is therefore very important. Stronger health systems are also key to making sure everyone has fair access to health in our connected world [2]. Effective systems are essential for handling health emergencies, such the COVID-19 pandemic, in which nations with more robust healthcare systems fared better. Investing in health systems benefits a nation's economy in addition to its citizens' health [3].

As healthcare increasingly relies on online methods such as remote doctor visits, new problems appear, especially concerning how data is kept safe. Protecting telehealth data is very important to make sure personal health details are secure, correct, and can only be seen by those who should see them [4]. Because telehealth is becoming a larger part of healthcare, good data safety measures are absolutely required. This keeps patients trusting the system and protects their privacy [5, 6]. Nevertheless, while technologies like encryption, user authentication, and access control are adequately put into consideration, telehealth systems still stand vulnerable under an ascendingly sophisticated and treacherous cyber-attack environment. Due to the increased value of medical information and also due to the interconnectedness of all telehealth platforms, they have now become prime targets for all sorts of cybercriminals. More often than not, hackers specifically target healthcare systems because of the immense value they place on personal and medical data-holding information ranging from basic demographic details and insurance info up until treatment histories and genetic data. This information has value for identity theft and financial fraud also could be breached and sold on the dark web, or even the Dark Web to make a huge profit. Ransomware attacks can stop telehealth services from working if important information is locked up and held for ransom. Even with security in place, patient care can be disrupted quickly, and important health information can be lost [7].

Because online dangers keep changing and targeting healthcare's online systems, especially telehealth, it's vital to always improve security. A data leak can do more than just lose important information. It can cause serious legal problems, harm a company's image, and most importantly, make patients lose faith in the safety of healthcare. Telehealth systems are at high risk because they often send and keep very private health information on many networks and devices. This increasing complexity requires a new way to protect health data, one that is better than old security methods. Blockchain is a promising answer, offering features like being decentralized, unchangeable, and open. These features can greatly improve the privacy, accuracy, and availability of patient data by offering records that can't be changed, tracking who accesses data, and storing data safely in many places. Looking into how blockchain can be used in healthcare can solve current security issues and lead to a more reliable, strong, and advanced digital health system.

The remainder of this paper is organized as follows. Section 2 provides an overview of the use of blockchain technology in telehealth and telemedicine, outlining its foundational concepts and potential benefits. Section 3 reviews related work, focusing on the evaluation of blockchain implementations within the healthcare sector. Section 4 presents the proposed methodology, describing the architecture of a blockchain-enabled telehealth system designed to enhance data integrity, privacy, and access control. Section 5 discusses the implementation results and assesses the system's effectiveness in managing healthcare data securely. Finally, Sect. 6 concludes the paper and outlines directions for future research.

2 Background

In 2008, blockchain technology was born with Satoshi Nakamoto's [8], introducing blockchain as the backbone of Bitcoin; the first-ever cryptocurrency. What started as a vision for decentralized digital money has since evolved into a powerful tool reshaping industries far beyond finance. Blockchain uses a shared record book that openly and permanently stores all transactions. Every block in the chain holds a set of confirmed transactions, and each block connects to the one before it, forming an ongoing chain. Figure 1 shows how blockchain works using a transaction between A and B as an example.

Blockchain tech, first made for digital money like Bitcoin, now does much more than just handle money. Its main qualities—being spread out, clear, and unchangeable—make things much safer, faster, and more reliable. Because of this, blockchain is changing many fields, especially.

Finance: Besides digital money, blockchain is changing regular money systems. It makes sending money across borders quicker and less expensive, simplifies how businesses trade, and allows people to own small pieces of valuable things. Smart contracts—agreements entered into the blockchain that run themselves—automate difficult financial chores, therefore saving money by eliminating the middleman.

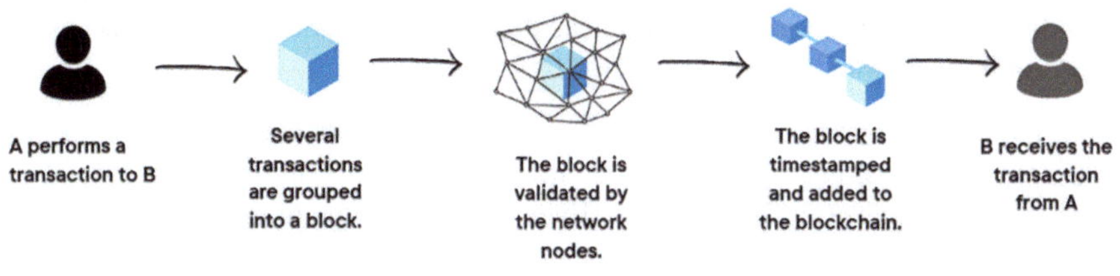

Fig. 1 Functioning of Blockchain technology

Supply Chain Management: Keeping track of products from where they start to when they're sold is hard. Blockchain offers a clear and unchangeable record that tracks every part of the supply chain. This makes sure products are real, shows where they came from, and proves they were made ethically. It gives better insight, lowers fraud, and makes shipping and delivery better.

Voting Systems: Because blockchain is clear and can't be changed, it's a good choice for safe and reliable voting systems.

Handling Ideas and Creations: Blockchain can help record and safeguard ownership of ideas, like patents, brand names, and copyrights. It creates a lasting and trustworthy record of who owns what, making it easier to manage permissions and fight against copying.

Managing Who You Are: Blockchain can give people more power over their personal information. It allows for safe and reliable online identities, which lowers the chance of identity theft and makes online activities easier.

Real Estate: Blockchain accomplishes this by managing financial transactions, ownership transfers, and property registration automatically. This results in enhanced process transparency, cost efficiency, and simplified record-keeping.

The main aspects of blockchain—its dispersion, transparency, and security—are what have made it growing in prominence in a range of domains. Healthcare is a field where blockchain has already shown good signs and has a lot of promise for the future.

Decentralization: Instead of having a main authority like a bank or government manage information, blockchain uses a network spread across many different people. Each participant, commonly known as a node, has a copy of the entire blockchain, so there is no single point of control. No central point of control implies that single points of failure come to an end, in addition to reducing susceptibility to manipulation or censorship. Decisions regarding the functioning of the network and its updates are made through consensus mechanisms; therefore, there is no possibility for one party to single-handedly determine the rules.

Transparency: Transparency is perhaps the most distinctive feature of blockchain. Users from any node on the network can view all the transactions and blocks made on the network. While this does not mean that the parties involved in the transaction will be identifiable, it does mean that most transactions are tied to anonymous pseudonymous addresses. This guarantees that all participants have equal access to required information.

Security: Blocks of transactions are kept safe with complex coding methods, especially hashing. Hashing turns transaction information into a short, unique code. This code is like a fingerprint for the block. Even a small change to the information will create a completely different code. Also, each block has the code from the block before it, making a linked and unchangeable list. This linking makes it very hard to change the blockchain.

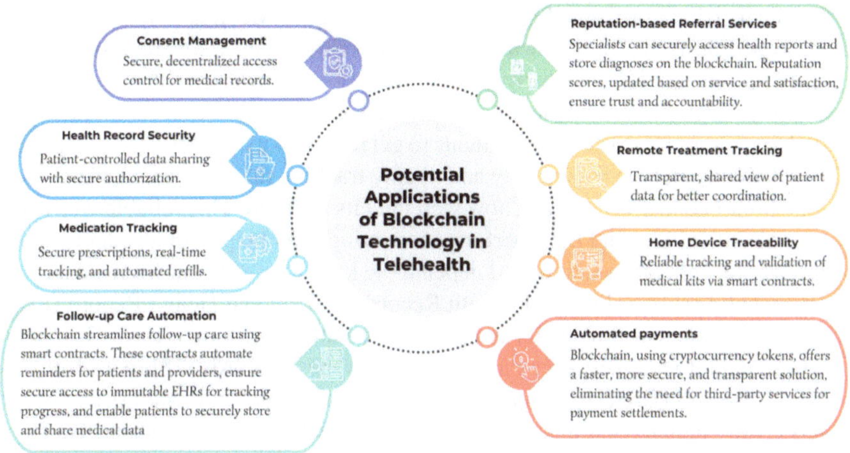

Fig. 2 Potential applications of blockchain technology in telehealth

With the growing interest in blockchain and its expanding range of applications, telehealth and telemedicine stand to gain significantly from its capabilities. Figure 2 offers a visual summary of these key opportunities, which we will now explore in greater detail.

- Automated Patient Consent Management: Electronic health records (HERs) contain very private details that must be safely shared between places like hospitals, drugstores, and health departments so patient information stays current. Telemedicine laws now let patients control their medical information by setting rules for how it's used and seen. Old ways of managing permission have problems, like slow sharing of records with experts, lack of trust in companies handling patient permissions, and difficulty in properly checking records [9, 10]. Blockchain can build trust because one of its key points, as we saw before, is security, which eliminates the need for a middleman.
- Tracking and Monitoring of Remote Treatments: Using technology to provide healthcare means patients and doctors need to connect online to properly check a person's health from a distance. This can happen in two main ways: directly between a patient and a doctor (D2C), or between healthcare businesses (B2B). In the first way, patients can talk to doctors about their health online. In the second way, caregivers can join online meetings for advice and training, like watching a surgery. During these online meetings, sharing videos and pictures, such as x-rays, helps doctors make correct diagnoses [11, 12]. However, current systems struggle to share patient records between different healthcare places. Blockchain can fix this by giving everyone involved a clear, shared view of a patient's records. This openness lets healthcare groups see a patient's past and plan the best treatment. For example, blockchain can show who looked at a patient's records and what they did with them.

- Secure Tracking of Home Medical Devices and Kits: Medical tools for home use allow people to check their own health outside of hospitals. Using readily available tests for self-checks and early disease finding can lower health expenses [13, 14]. However, current telehealth systems lack clear information about these kits, making it hard for doctors and patients to get reliable kits from trusted companies. Blockchain can help by securely and openly tracking the buying and use of these kits. Smart contracts can track how well the kits perform. This reliable data about home medical kits and their performance ratings can help patients, doctors, and lab workers choose accurate and dependable kits from well-known makers.
- Protected Access to Personal Health Records: A Personal Health Record (PHR) holds a person's health details and other information about their care. The person whose data it is makes, keeps up, and handles these records [15]. An Electronic Health Record (EHR) has more complete health records, and doctors and hospitals create, update, and manage them. Most virtual healthcare systems use cloud platforms, which can be risky because one organization controls them. This can also make PHRs less secure. Blockchain technology helps patients keep their data private. Smart contracts can sign up and allow users to see patient data based on what the patient allows. Blockchain also lets patients easily share and manage their data with authorized people, following the rules they set [16–18].
- Reliable Tracking of Medication Delivery and Pharmacy Refills: Healthcare systems using online doctor visits need doctors to use blockchain to send prescriptions to pharmacies. Blockchain's security features help prevent mistakes and changes to prescriptions [19, 20]. Pharmacists can then check the prescription on the blockchain, prepare the medicine, and send it to patients. Delivery services can update the blockchain with the package's location so pharmacists and patients can follow the delivery. The transparency and traceability of blockchain makes it easier for patients and doctors to ensure the medicine is real by checking its history [21]. Smart contracts can automatically order refills when needed. The pharmacy then confirms the refill, sends the medicine, and updates the records.

3 Related Work

Blockchain technology has gained increasing attention in the health sector as a potential answer to the long-term issues of data security, interoperability, and trust in health systems. Increasing numbers of studies have evaluated its effectiveness, particularly in guaranteeing privacy and increasing the integrity of electronic health records (EHRs). For instance, MedRec [22] is a blockchain system that seeks to revolutionize the management of medical records by prioritizing patient agency and offering immutable, tamper-evident audit trails. In contrast to traditional, typically fragmented, methods of storing and sharing sensitive health data, MedRec leverages the inherent security and transparency of blockchain technology to allow patients to have greater control over their own data. The system imagines a world where patients could simply give or revoke access to their records by various healthcare providers, opening up healthcare to greater collaboration and effectiveness. Their initial prototype was a success in putting the promise of blockchain into practice and creating an open and decentralized data-sharing infrastructure among various healthcare providers. This proof-of-concept illustrated the ability for greater interoperability, reduced administrative overhead, and improved data security compared to existing centralized systems. Also, the tamper-proof audit trails offered by the blockchain enable a clear and traceable record of all access to and modification of data that is conducive to trust and accountability in the healthcare network. This method not only enhances patient privacy but also reduces the risk of data breaches and illegal alterations, and ultimately results in a safer and more reliable healthcare system.

In the same manner, FHIRChain [7], an architecture that utilizes blockchain technology in harmony with HL7 FHIR standards, and allows for secure, scalable sharing of clinical data. This architecture was tested for conformity to HIPAA privacy regulations as well as preserving data provenance. In a broader review, Agbo, Mahmoud, and Eklund [23] systematically reviewed blockchain uses in healthcare and concluded that although the technology has potential in applications such as health data management, drug traceability, and integrity of clinical trials, problems remain as much regarding scalability, admissibility to courts, and integration into existing systems [24]. These evaluations underscore an important fact too readily overlooked in the hype surrounding blockchain technology: although blockchain is no panacea or "magic bullet" that can solve all of the problems facing healthcare, its thoughtful constructed and strategic use in specific healthcare environments holds much promise. This commitment is based primarily on its potential to improve data governance significantly, raise transparency, and build a greater level of trust between patients, providers, and other stakeholders [25, 26]. For example, using blockchain's built-in security and shared record-keeping, hospitals and clinics can better protect patient information when it's viewed or changed. This increased responsibility helps stop unauthorized access and keeps data correct. Also, because blockchain is open, people can better understand where information comes from and how it's used. This gives patients more knowledge about their data and possibly more control over who sees it. In the end, blockchain's better transparency and safety can foster greater confidence, leading patients to be more open to sharing vital health information and engaging in their treatment. But for success, careful planning, a clear understanding of its applications, and awareness of its limitations are essential. These limits include how well it grows, how it works with current systems, and the need for rules to manage its decentralized nature.

4 Methodology

Telehealth systems handle very private health information like patient records, test results, and care plans. Keeping this information safe, private, and easy to get to is extremely important. Only people allowed, like doctors, nurses, and patients, should be able to see it. Blockchain technology can help with this. We will talk about the different ways blockchain can protect telehealth data. We will look at how blockchain can manage patient IDs, control who sees medical records, protect data when it is sent, and make sure medical devices are safe. Blockchain can not only make things more secure but also make systems work better together, speed up processes, and give patients more control over their health information.

- Use ledge. This ledge must keep track of all activity in a blockchain. This ledge cannot be changed (immutable). Old information stays as it is, never erased or altered. In blockchain, these ledges are spread out. This means no single person can get into the system to look at transactions or private details. People are only allowed to see the information stored in the ledger.
- In a blockchain, using a chain of blocks is one of the key security mechanisms. Each block must have a unique code called a hash. Blocks are linked together using the hash of the previous block. If someone tries to alter the data in a block, its hash changes, which in turn affects the entire chain. This makes it more difficult to tamper with important data, thereby enhancing security [27].
- Using smart contract smarts in blockchain makes decentralized apps and systems much stronger and reliable. Contract smarts means adding clever rules and automatic checks into smart contracts. This helps find and stop weaknesses that bad people could use. This includes checking for possible attacks early, using automatic tests to meet safety rules, and watching for strange activity to fix problems right away [28].
- Blockchain is a system that doesn't rely on a central authority. It mostly uses direct communication between users (peer to peer communication). Each computer in the network is called a node. Thousands of these nodes have a copy of the shared record. This record is used to verify transactions. If a node disagrees with a transaction, it won't go through and will be stopped. This helps prevent dishonest transactions [29].

With its emphasis on strong security measures as described above, blockchain technology offers a compelling and promising approach to building a healthcare ecosystem that is not only more trustworthy and transparent, but also significantly more cost-effective. Figure 3 shows the architectural design of a telemedicine ecosystem using blockchain technology. The architecture shows how blockchain can be integrated into various telemedicine components to improve security, increase data integrity, and enable seamless communication between patients, doctors, and other stakeholders in the healthcare network.

This system uses blockchain to make health data more secure and easier to manage, allowing patients, devices, and health organizations to share information. It also uses decentralized storage, like IPFS, to manage and access large files such as medical images and patient records.

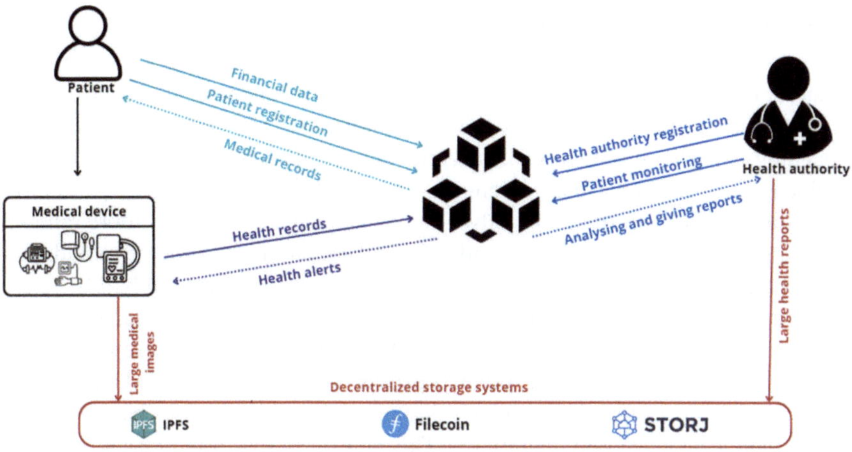

Fig. 3 Blockchain-powered telemedicine ecosystem design

5 Results

The system in Fig. 3 uses blockchain to make health data safer and easier to control. Using a shared, unchangeable record, the system keeps patient details safe from those who shouldn't see them and prevents changes without permission. This improved safety builds more confidence between patients, devices, and healthcare groups, which helps them share important information more easily. Instead of using central databases, which can easily fail, patients have more say over their own data. They can choose who gets to see their information, like doctors or research groups.

To handle storing and finding big, messy data sets well, the system uses storage that's spread out, like the Inter Planetary File System (IPFS). Normal storage that's all-in-one place can be expensive and slow for huge files like detailed medical pictures (X-rays, MRIs, and CT scans), full patient records, and gene information. IPFS is cheaper and can grow easily because it spreads files across many computers, making sure data isn't lost, stays correct, and is found quickly. When someone needs a file, the system gets it from the closest IPFS computer, making it faster and more dependable for the user.

This combination of blockchain for security and IPFS for efficient storage marks is a big step forward in how healthcare information is handled. This opens doors for healthcare systems that work together better, focus on patients, and use data to make smart choices.

6 Conclusion and Future Works

In short, using blockchain with telehealth can really improve healthcare. It can boost security, clarity, and efficiency, fix current issues. Telehealth, which is rapidly expanding, often has problems with data protection, system compatibility, and patient confidence. Blockchain's main strengths—its permanent records, strong security, and ability to share information easily—can make telehealth systems more reliable, better at protecting patient privacy, and safer and simpler for healthcare providers to collaborate. Because blockchain records cannot be secretly altered, it builds a trustworthy and provable history of patient treatment. This is very important in telehealth, where confirming the accuracy and integrity of shared information is essential. Strong security, such as complex coding, protects patient data from hackers and other online threats, which are increasing in today's digital world. This provides better security against data leaks from one main place. This control is very important to get more people to use and believe in telehealth, letting them take charge of their health.

Future studies will look more at how AI and blockchain can work together to make health data management better in real-time and find new ways to diagnose and treat illnesses. For instance, AI programs could be added to blockchain systems to find diseases early by carefully studying genetic data, medical pictures, and other important health information. By safely keeping and sharing this data on a blockchain, scientists can speed up the creation of personalized treatments and make patients healthier. Imagine AI programs looking at a patient's chances of getting a certain disease, with the results safely saved on a blockchain to make sure they are correct and can't be changed. This information can then be used to manage the patient's health early on and maybe stop the disease from starting.

References

1. World Health Organization, (2010). World health statistics 2010. World Health Organization.
2. Kruk, M.E., Gage, A.D., Joseph, N.T., Danaei, G., García-Saisó, S. and Salomon, J.A., (2018). Mortality due to low-quality health systems in the universal health coverage era: a systematic analysis of amenable deaths in 137 countries. The Lancet, 392(10160), pp.2203-2212.
3. Thomas, S., Sagan, A., Larkin, J., Cylus, J., Figueras, J. and Karanikolos, M., (2020). Strengthening health systems resilience: key concepts and strategies.
4. Kruse, C.S., Krowski, N., Rodriguez, B., Tran, L., Vela, J. and Brooks, M., (2017). Telehealth and patient satisfaction: a systematic review and narrative analysis. BMJ open, 7(8), p.e016242.
5. Aljedaani, B. and Babar, M.A., (2021). Challenges with developing secure mobile health applications: systematic review. JMIR mHealth and uHealth, 9(6), p.e15654.
6. Iwaya, L.H., Ahmad, A. and Babar, M.A., (2020). Security and privacy for mHealth and uHealth systems: a systematic mapping study. IEEE Access, 8, pp.150081-150112.
7. Zhang, P., White, J., Schmidt, D.C., Lenz, G. and Rosenbloom, S.T., (2018). FHIRChain: applying blockchain to securely and scalably share clinical data. Computational and structural biotechnology journal, 16, pp.267-278.
8. Berentsen, A. (2019). Aleksander berentsen recommends "bitcoin: a peer-to-peer electronic cash system" by Satoshi Nakamoto. 21st Century economics: Economic ideas you should read and remember, 7–8.

9. Dubovitskaya A., Xu Z., Ryu S., Schumacher M., Wang F, (2017) "Secure and trustable electronic medical records sharing using blockchain," American Medical Informatics Association, AMIA annual symposium proceedings.
10. E. Saweros, Y.-T. Song, (2019) "Connecting heterogeneous electronic health record systems using tangle," International Conference on Ubiquitous Information Management and Communication, Springer, Phuket, Thailand.
11. K. Mannaro, G. Baralla, A. Pinna, S. Ibba, (2018) "A blockchain approach applied to a teledermatology platform in the Sardinian region (Italy)," Information 9.
12. J. Gorke, (2020) "Telehealth Continues To Change The Face Of Healthcare Delivery," Online]. Accessed on 27.04 (2020).
13. Li, Raymond. "Multifunctional self-diagnostic device for in-home health-checkup." U.S. Patent Application No. 10/904,818.
14. S.M. Weissman, K. Zellmer, N. Gill, D. Wham, (2018) " Implementing a virtual health telemedicine program in a community setting, J. Genet. Counsel".
15. M. Fetter, (2008) "Personal health records: protecting behavioral health consumers," rights. Issues in mental health nursing, pp. 720–722.
16. G.G. Dagher, J. Mohler, M. Milojkovic, P.B. Marella, (2018) "Ancile: privacy-preserving framework for access control and interoperability of electronic health records using blockchain technology," Sustain. Cities Soc., p. 283–297.
17. R. Guo, H. Shi, D. Zheng, C. Jing, C. Zhuang, Z. Wang, (2019) "Flexible and efficient blockchain-based ABE scheme with multi-authority for medical on demand in telemedicine system," IEEE Access 7, p. 88012–88025.
18. A.F. da Conceiçao, F.S.C. da Silva, V. Rocha, A. Locoro, J.M. Barguil, (2018) "Eletronic Health Records Using Blockchain Technology," arXiv preprint arXiv:1804.10078.
19. G. Pallotta, (2021) "Development of Telehealth systems and technologies for remote patients' healthcare".
20. Y. El-Miedany, (2017) "Telehealth and telemedicine: how the digital era is changing standard health care," Smart Homecare Technol. Telehealth 4, p. 43–51.
21. M. Thakore, (2020) "Transforming Healthcare: Blockchain Based Medical Prescription," 2020.
22. Azaria, A., Ekblaw, A., Vieira, T. and Lippman, A., (2016), August. Medrec: Using blockchain for medical data access and permission management. In 2016 2nd international conference on open and big data (OBD) (pp. 25–30). IEEE.
23. Agbo, C.C., Mahmoud, Q.H. and Eklund, J.M., (2019), April. Blockchain technology in healthcare: a systematic review. In Healthcare (Vol. 7, No. 2, p. 56). MDPI.
24. Roehrs, A., Da Costa, C.A., da Rosa Righi, R. and De Oliveira, K.S.F., (2017). Personal health records: a systematic literature review. Journal of medical Internet research, 19(1), p.e5876.
25. Angraal, S., Krumholz, H.M. and Schulz, W.L., (2017). Blockchain technology: applications in health care. Circulation: Cardiovascular quality and outcomes, 10(9), p.e003800.
26. Griggs, K.N., Ossipova, O., Kohlios, C.P., Baccarini, A.N., Howson, E.A. and Hayajneh, T., (2018). Healthcare blockchain system using smart contracts for secure automated remote patient monitoring. Journal of medical systems, 42, pp.1-7.
27. Zheng, Z., Xie, S., Dai, H., Chen, X., & Wang, H. (2017). An Overview of Blockchain Technology: Architecture, Consensus, and Future Trends. Proceedings of the 2017 IEEE International Congress on Big Data (BigData Congress), 557–564.
28. Atzei, N., Bartoletti, M., & Cimoli, T. (2017). A Survey of Attacks on Ethereum Smart Contracts (SoK). Proceedings of the 6th International Conference on Principles of Security and Trust (POST), 164–186.
29. Tschorsch, F., & Scheuermann, B. (2016). Bitcoin and Beyond: A Technical Survey on Decentralized Digital Currencies. IEEE Communications Surveys & Tutorials, 18(3), 2084-2123.
30. Attou, H., Guezzaz, A., Benkirane, S. and Azrour, M., 2025. A New Secure Model for Cloud Environments Using RBFNN and AdaBoost. SN Computer Science, 6(2), p.188.
31. Mohy-eddine, M., Guezzaz, A., Benkirane, S. and Azrour, M., 2024. Malicious detection model with artificial neural network in IoT-based smart farming security. Cluster Computing, 27(6), pp.7307-7322.

32. Dargaoui, S., Azrour, M., El Allaoui, A., Guezzaz, A. and Benkirane, S., 2025. Machine learning for security boosting in Internet of Things environments. In Recent Advances in Internet of Things Security (pp. 63–71). CRC Press.
33. Dargaoui, S., Azrour, M., El Allaoui, A., Guezzaz, A., Alabdulatif, A. and Ahmad, S., 2025. ECC-Based Anonymous and Multi-factor Authentication Scheme for IoT Environment. International Journal of Online & Biomedical Engineering, 21(1).
34. Blali, A., Dargaoui, S., Azrour, M., Guezzaz, A., Amounas, F. and Alabdulatif, A., 2025. Analysis of deep learning-based intrusion detection systems in IoT environments. EDPACS, pp.1–35.
35. Hazman, C., Guezzaz, A., Benkirane, S. and Azrour, M., 2025. A smart model integrating LSTM and XGBoost for improving IoT-enabled smart cities security. Cluster Computing, 28(1), p.70.
36. Mohy-Eddine, M., Guezzaz, A., Benkirane, S., Azrour, M. and Bella, K., 2025. A monitoring system with deep learning for IoT smart environments security. In Recent Advances in Internet of Things Security (pp. 93–104). CRC Press.
37. Dargaoui, Souhayla, Mourade Azrour, Ahmad El Allaoui, Azidine Guezzaz, Said Benkirane, Abdulatif Alabdulatif, and Fatima Amounas. "Internet-of-Things-Enabled Smart Agriculture: Security Enhancement Approaches." In 2024 4th International Conference on Innovative Research in Applied Science, Engineering and Technology (IRASET), pp. 1–5. IEEE, 2024.
38. Anwar, T., Khan, G.A., Ashraf, Z., Ansari, Z.A., Ahmed, R. and Azrour, M., 2024. The combination of blockchain and the internet of things (iot): applications, opportunities, and challenges for industry. Blockchain and Machine Learning for IoT Security, pp.56–76.
39. Assmi, H., Guezzaz, A., Benkirane, S., Azrour, M., Jabbour, S., Innab, N. and Alabdulatif, A., 2025. A Robust Security Detection Strategy for Next Generation IoT Networks. Computers, Materials & Continua, 82(1).

Integrating Transfer Learning with Fingerprint Recognition for Enhanced Student Attendance Systems

Slimane Ennajar, Walid Bouarifi, and Khalid Aitbenhamou

Abstract This study investigates the integration of transfer learning methods with fingerprint recognition technologies to enhance student attendance systems, providing a comparative analysis between two pre-trained models: ResNet50 and MobileNetV2. Given the critical importance of accurately and efficiently identifying students in educational settings, adopting advanced neural network architectures offers a promising avenue for improving the functionality and design of attendance systems. Our methodology involves fine-tuning these models on specific datasets of fingerprint images and assessing their performance using key metrics: accuracy, precision, recall, F1 score, and the Matthews Correlation Coefficient (MCC). The comparative analysis revealed that the ResNet50 model outperformed MobileNetV2 with an impressive 100% accuracy rate, while MobileNetV2 also demonstrated significant effectiveness, achieving an accuracy of 99.66%. This study underscores the potential of employing advanced neural networks through transfer learning to bolster educational attendance systems.

Keywords Transfer learning · Student attendance system · Fingerprint recognition · ResNet50 · MobileNetV2

S. Ennajar (✉) · W. Bouarifi
Mathematical Team and Information Processing National School of Applied Sciences SAFI, Cadi AYYAD University, Marrakech, Morocco
e-mail: Slimane.ennajar@ced.uca.ma

K. Aitbenhamou
LISI Laboratory, Faculty of Sciences, University of Cadi Ayyad, Marrakech, Morocco
e-mail: khalid.aitbenhamou@ced.uca.ma

M. Azrour and A. Alabdulatif (eds.), *AI-Driven Security for Next-Generation IoT Systems*, https://doi.org/10.1007/978-3-032-08784-3_13

1 Introduction

In this era of digital transformation, educational institutions are continuously seeking innovative technologies to streamline business processes and enhance security measures. A key area of focus is the attendance tracking system, a crucial component for monitoring student engagement and academic performance. However, traditional attendance methods are often marred by inefficiencies such as time consumption, vulnerability to fraud, and the absence of real-time analytics. Advanced deep learning models, particularly ResNet50 and MobileNetV2 [1, 2], offer significant advantages in addressing these issues, thanks to their capacity for more efficient and secure processing of data. These models herald a new phase of enhanced reliability and accuracy in attendance systems.

ResNet50 and MobileNetV2 were selected for comparative analysis due to their unique architectural characteristics and performance efficiency in image recognition tasks, including biometric recognition [3]. Known for its deep residual networks, ResNet50 excels at handling complex image data and extracting high-level features, making it ideal for intricate classification tasks [4, 5]. Contrary to what one might assume, MobileNetV2, which is designed for mobile devices, provides a balance between accuracy and computational efficiency. This balance is critical for implementing scalable attendance systems in educational environments with limited resources.

This paper contributes to the burgeoning field of educational technology by demonstrating how Transfer Learning [6], coupled with state-of-the-art fingerprint recognition algorithms [7, 8], can address the current limitations of student attendance systems. Through comprehensive comparative analysis, it aims to provide insights into the practicality, scalability, and effectiveness of these models, offering a path forward for educational institutions eager to adopt more sophisticated, reliable, and cost-effective attendance tracking solutions.

Our research utilizes ResNet50 and MobileNetV2 pre-trained models to enhance fingerprint-based student attendance tracking. The main contributions and innovations of our work include:

- Adapting transfer learning strategies for student attendance systems with established models such as ResNet50 and MobileNetV2.
- Implementing data augmentation techniques to address the challenge of limited dataset images, ensuring robust validation of our approach.
- Conducting a comprehensive validation of our method by evaluating various performance indicators, including Accuracy, Precision, Recall, F1 Score, and MCC.
- Achieving exceptional results, demonstrating a perfect 100% accuracy rate in fingerprint recognition for student attendance tracking.

The rest of this document is organized in the following manner: Sect. 2 offers a review of prior research concerning fingerprint recognition and transfer learning methods. Section 3 outlines the materials and methodologies employed in this study.

Detailed experimental findings are disclosed in Sect. 4, and Sect. 5 concludes the paper.

2 Related Work

In the study [6], the authors a thorough review of transfer learning, emphasizing its role in utilizing knowledge from one area to enhance learning in another, particularly when there is a lack of data. It categorizes and examines over forty methods of homogeneous transfer learning, considers its various applications, and includes experiments comparing over twenty models across three datasets to demonstrate the effectiveness of different approaches. In [9], the authors discuss the advantages of using fingerprint recognition due to its reliability and cost-effectiveness. They detail a system that employs fingerprint scanning technology for precise identity verification based on unique fingertip patterns. This system can function independently or alongside a computer, and is designed to automate attendance tracking, enhance security, and prevent fraud. The primary goal of this approach is to streamline administrative processes and maintain accurate attendance records. In [10], the authors proposed a system based on fingerprint recognition technology for managing student attendance in educational institutions. The system's effectiveness was validated through a series of tests involving the enrollment and authentication of 15 students' fingerprints, yielding remarkable results: a 0% False Acceptance Rate (FAR), a 100% True Accept Rate (TAR), and a 0% False Reject Rate (FRR). These metrics underscore the system's potential to significantly reduce fraud, thereby offering a highly reliable and error-free approach to student attendance management. In [11], the authors introduce a novel method combining transfer learning with fingerprint biometrics to enhance face recognition, crucial for forensic investigations. By leveraging Bregman divergence regularization, their approach effectively reduces the distribution differences between the fingerprint and facial image domains, facilitating the identification of a common subspace for improved classification under varying conditions. In [12], the authors propose a novel attendance management system designed to monitor student presence in educational settings and communicate this information to their parents. The system utilizes fingerprint recognition technology to ensure a reliable and tamper-proof method of marking attendance. A key feature of this system is its ability to send SMS notifications to parents regarding their child's attendance status, leveraging GSM technology for communication. In [13], the authors explored the development of a fingerprint-based attendance system aimed at improving the process of recording student attendance at Kolej University Poly-Tech Mara. This system, designed to address the inefficiencies and potential for fraud inherent in manual attendance tracking, utilizes fingerprint biometrics to provide a reliable and secure method of student identification. The study leverages evolutionary prototyping as its methodology, enabling iterative feedback and refinement to enhance the system's effectiveness. The implementation facilitates more systematic, efficient, and ethical monitoring of student attendance, showcasing the potential of biometric technologies in educational administration. The authors, in [14] utilized deep network models for classifying new sets of fingerprint images without the need for retraining from scratch. By leveraging real datasets from Can Tho University students, the study trained models such as VGG, ResNet50, Inception-v3, and Xception on 9236 fingerprint images from 441 students. This innovative approach involves using these models as feature extractors and fine-tuning only the last layer for new datasets. Empirical results from three fingerprint image datasets (FP-235, FP-389, and FP-559) showed that the deep network models achieved a minimum accuracy of 96.72% on the test sets, with the ResNet50 models achieving classification accuracies of 99.00%, 98.33%, and 98.05% on FP-235, FP-389, and FP-559, respectively. In [15], the authors developed an attendance management system that leverages fingerprint biometrics. This system aims to automate the process of recording attendance in educational settings, enhancing efficiency and security while reducing the potential for fraudulent attendance marking. It eliminates the need for manual intervention by educators during lectures, automates attendance management using pre-set schedules, and enables the automatic evaluation of students' attendance. Furthermore, the system provides real-time online access for guardians to monitor their wards' attendance and receive reports via email.

3 Materials and Methods

3.1 Dataset

The CASIA Fingerprint Database, developed by the Institute of Automation at the Chinese Academy of Sciences, serves as a pivotal resource for biometric research, particularly in fingerprint recognition. It features a comprehensive collection of high-quality fingerprint images from multiple subjects, designed to encompass a wide range of variations in finger position, pressure, and rotation. This diversity is crucial for developing and evaluating fingerprint recognition algorithms, offering researchers the ability to test their systems under varied and realistic conditions. The database is valued for its high-resolution images, a substantial volume of data, and the inclusion of intra-class variations, making it an indispensable tool for those aiming to enhance the accuracy and reliability of fingerprint recognition technologies [16].

In our study, we employed a dataset obtained from CASIA Fingerprint V5, which includes 10,380 fingerprint images across 10 different classes, illustrated in Fig. 1. The evaluation process involved validation and test sets from new databases, containing 1662 and 2071 images, respectively. The structure of this dataset is visually outlined in Fig. 2.

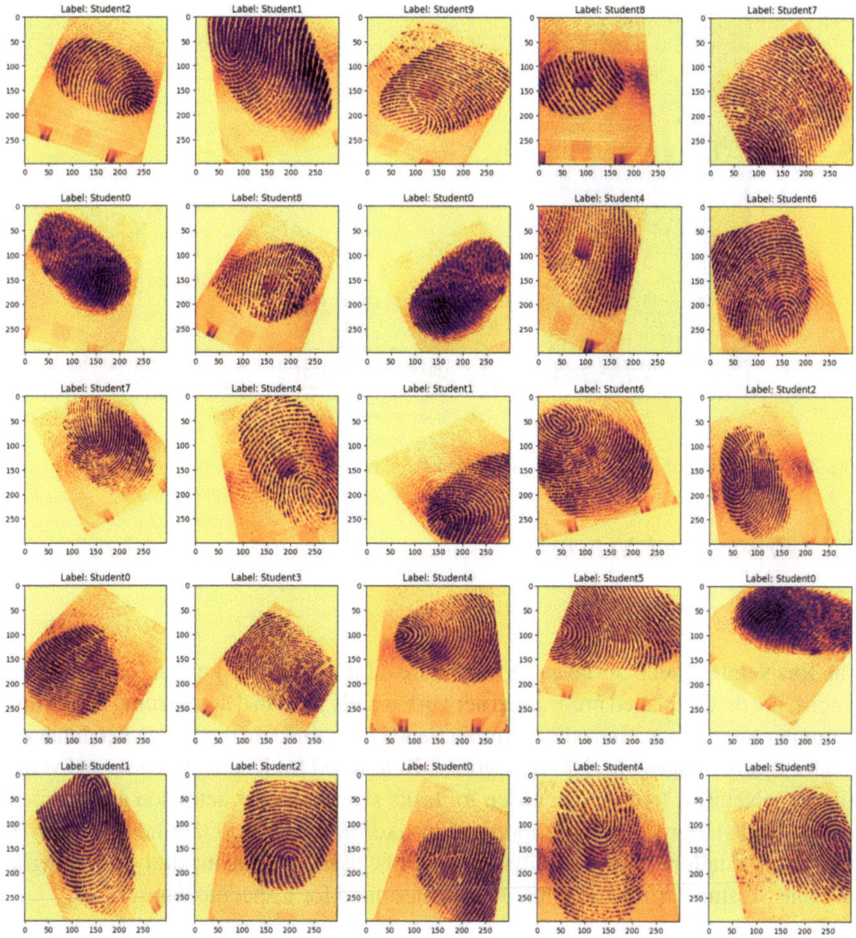

Fig. 1 Variations of fingerprint patterns: a visual collection from CASIA dataset

3.2 Data Augmentation Technique

In our work, we applied data augmentation techniques to artificially expand our dataset and improve the robustness of our fingerprint recognition models. We introduced geometric variations by rotating, flipping, and scaling the images, and we adjusted photometric properties by altering brightness and contrast to mimic different lighting conditions. Noise was added to simulate imperfections in fingerprint scans, and random erasing was employed to ensure the model's resilience to partial occlusions. Additionally, we utilized elastic transformations to account for the natural distortions of fingerprints on varied surfaces. These methods effectively increased the diversity of our training data, enabling the development of a more accurate and reliable attendance system.

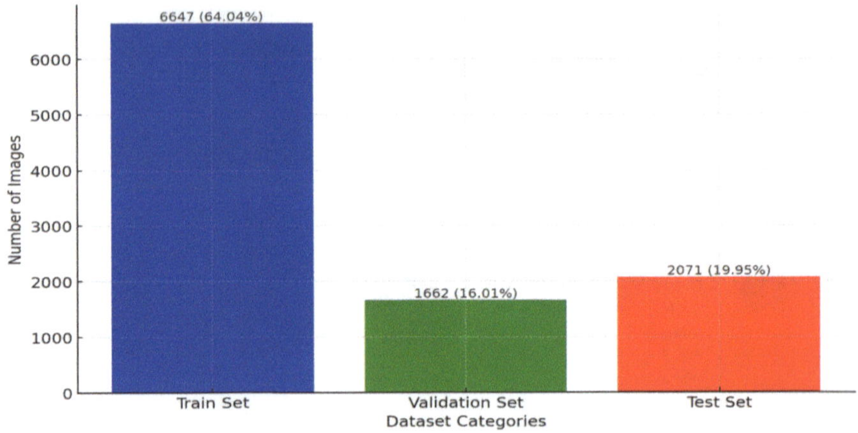

Fig. 2 Distribution of dataset images with counts and percentages

3.3 Pre-trained Networks

ResNet50

The ResNet50, a deep 50-layer version of Residual Networks known for its residual blocks, is a distinguished pre-trained network widely used in deep learning for various computer vision tasks. Introduced in 2015 and pre-trained on the ImageNet database, it can classify images into 1000 categories. Renowned for its effectiveness and versatility, ResNet50 is frequently chosen for tasks such as object detection and segmentation [17, 18]. For specialized applications like student group classification, its final layers can be tailored to the required number of classes—10 in our study—leveraging its robust feature extraction while being fine-tuned for a specific dataset.

MobileNetV2

MobileNetV2 is a compact, efficient convolutional neural network architecture tailored for mobile and edge devices [2, 19]. Introduced by Google researchers in 2018, it improves upon its predecessor, MobileNetV1, by incorporating an inverted residual structure with shortcut connections between the bottleneck layers. This design reduces the model's size and computational complexity, enabling it to run swiftly on devices with limited processing power and memory without significantly sacrificing performance. MobileNetV2 is adept at a wide range of tasks, including image.

3.4 *Parameter Setting*

The parameter settings for training transfer learning models focus on two different configurations to highlight variations in training dynamics. Both configurations use an input shape of (299, 299, 3), which is typical for image processing models requiring three color channels. The number of epochs varies between the setups, with one training for 30 epochs and the other for 40 epochs, illustrating an experiment to determine the effect of training duration on model performance. The learning rate is set at a low value of 0.0001 for both configurations, utilizing the Adam optimizer to enhance the convergence rate. The activation function used is SoftMax, which is suitable for multi-class classification tasks and indicative of a scenario with 10 distinct classes, such as student identification or similar categorization tasks.

4 Results and Discussions

4.1 *Evaluation Metrics*

Table 1 presents the performance metrics of ResNet50 and MobileNetV2 models, analyzed at 30 and 40 epochs. At 30 epochs, ResNet50 achieves perfect scores across all metrics [20, 21]: Accuracy, Precision, Recall, F1 Score, and MCC, all at 100%. MobileNetV2, while highly effective, scores slightly lower, with an Accuracy, Precision, Recall, and F1 Score of 99.66% and an MCC of 99.63%. When evaluated at 40 epochs, ResNet50 maintains near-perfect performance with all metrics at 99.95%. MobileNetV2 shows a decline compared to its 30-epoch performance, with Accuracy, Precision, Recall, and F1 Score dropping to around 98.31% and MCC [20] to 98.13%. These results indicate that ResNet50 consistently outperforms MobileNetV2 across all considered metrics, maintaining higher stability in performance as the number of epochs increases. This comparison highlights the effectiveness and reliability of ResNet50 in achieving high precision across various performance metrics over multiple training cycles.

Table 1 Performance metric

Model	Number of epochs	Accuracy score (%)	Precision score (%)	Recall score (%)	F1 score (%)	MCC (%)
ResNet50	30	100	100	100	100	100
MobileNetV2	30	99.66	99.67	99.66	99.66	99.63
ResNet50	40	99.95	99.95	99.95	99.95	99.95
MobileNetV2	40	98.31	98.42	98.31	98.31	98.13

4.2 Loss and Accuracy

Figure 3 displays two-line graphs detailing the training and validation of a ResNet50 model across 30 epochs. On the left, the graph depicts the training and validation loss, with both decreasing sharply initially and then fluctuating before stabilizing as the epochs progress. On the right, the graph shows the training and validation accuracy, which increases rapidly at first and then plateaus, with the validation accuracy demonstrating some variance. This suggests the model is learning effectively, with some signs of overfitting or instability in later epochs as indicated by the fluctuations in loss and accuracy.

Figure 4 displays the evolution of the MobileNetV2 model's performance during a 30-epoch training period. The loss graph on the left reveals a sharp initial decline in both training and validation loss, followed by a phase of lower variability, suggesting early learning and later stabilization. The accuracy graph on the right demonstrates a rapid ascent in training accuracy that quickly reaches a high plateau, while the validation accuracy exhibits a more gradual increase with noticeable fluctuations. This performance profile is typical of neural network training dynamics, reflecting both learning efficiency and the challenges of model generalization.

Figure 5 presents the training and validation phases of a ResNet50 model over 40 epochs. The left graph indicates the loss, where both training and validation loss quickly decrease and then level off, showing the model's learning stability. The right graph tracks accuracy, with the training accuracy achieving high-levels early on and maintaining a plateau, while validation accuracy also climbs to a high plateau but with some notable variance. This variance in validation accuracy could prompt further fine-tuning to ensure the model generalizes well to new data.

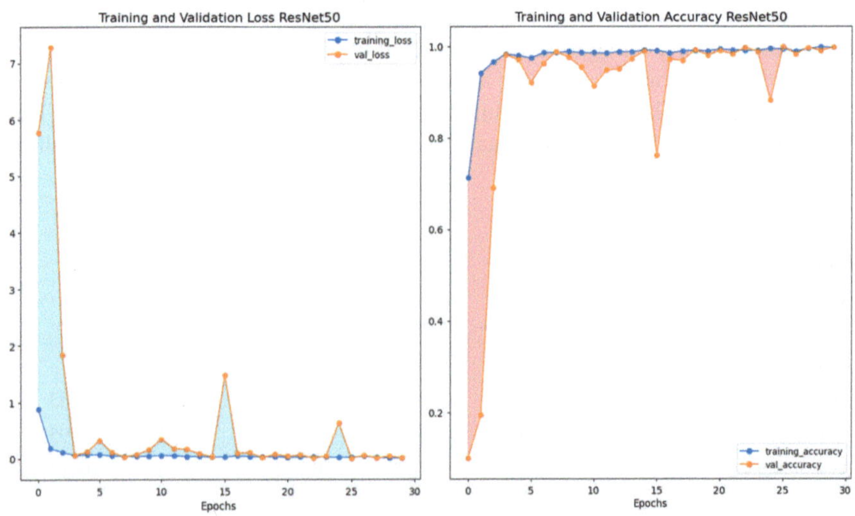

Fig. 3 Training and validation of ResNet50 model with 30 epochs

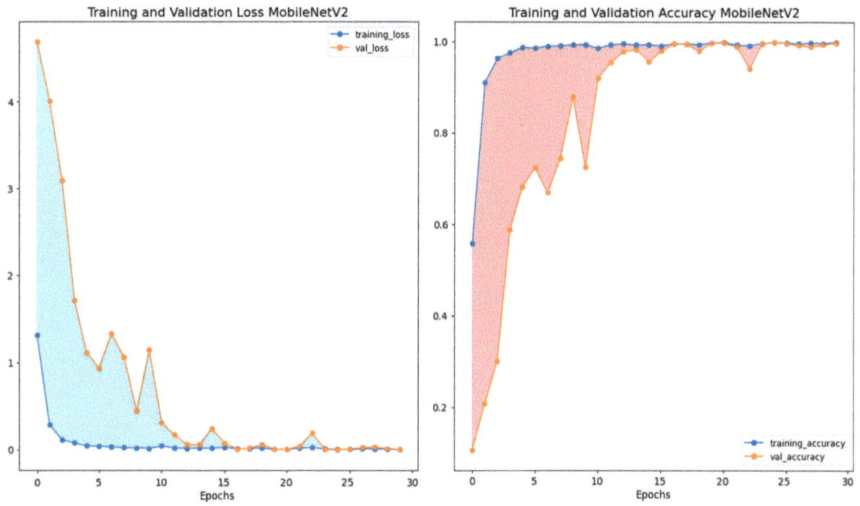

Fig. 4 Training and validation of MobileNetV2 model with 30 epochs

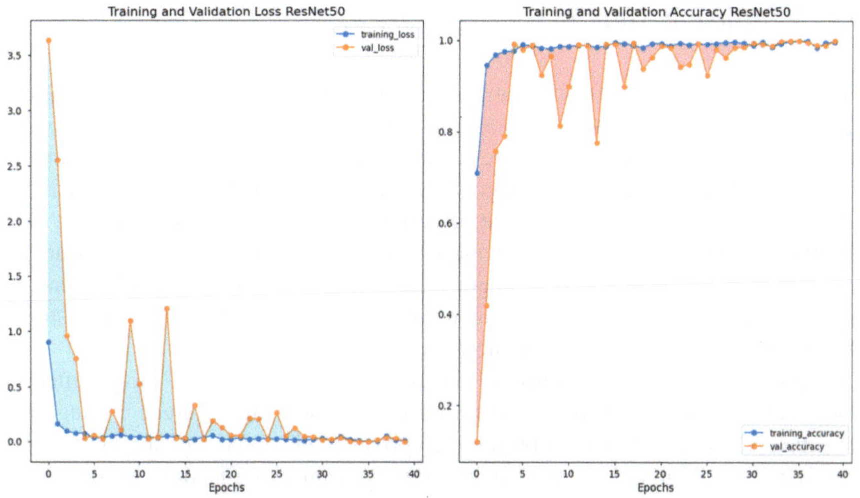

Fig. 5 Training and validation of ResNet50 model with 40 epochs

Figure 6 shows the training of the MobileNetV2 model across 40 epochs, where loss metrics rapidly decrease and stabilize, and accuracy metrics quickly rise and plateau, with some fluctuations in validation accuracy indicating variations in model performance on new data.

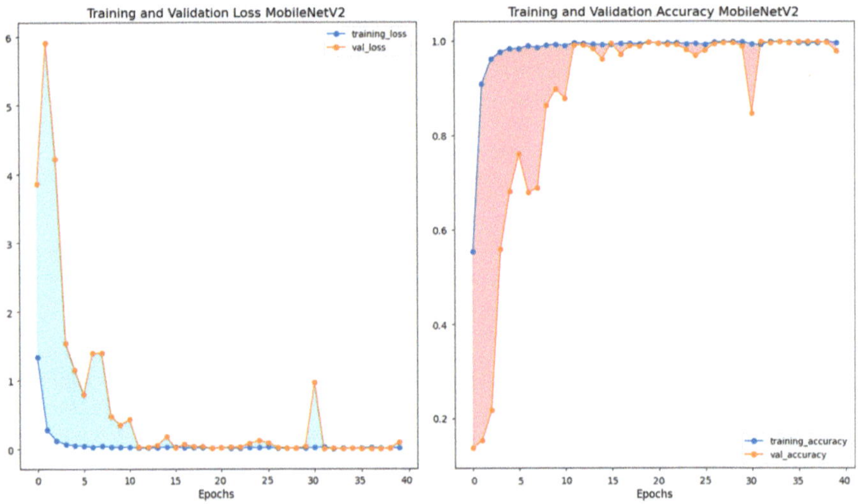

Fig. 6 Training and validation of MobileNetV2 model with 40 epochs

4.3 Confusion Matrix

Figure 7 displays the confusion matrix for the ResNet50 model, illustrating classification accuracy across ten student categories after training for a specified number of epochs. Each row corresponds to the actual classes of students, and each column to the predicted classes. The main diagonal, where the true labels match the predicted labels, shows high values indicative of correct classifications; for example, "Student00" has 207 instances correctly classified. This pattern of accuracy repeats for all categories, indicating strong diagonal dominance and no misclassifications, as evidenced by the absence of off-diagonal numbers above zero. This demonstrates the model's ideal and highly accurate performance.

Figure 8 showcases a confusion matrix for the MobileNetV2 model after 30 epochs of training, displaying its performance in classifying ten student categories. The matrix's diagonal cells, representing correct predictions, indicate high values for each class, though there are a few instances of misclassification, as seen in the non-zero off-diagonal cells. These results reflect the model's overall high accuracy with slight discrepancies in certain classes.

Figure 9 displays the confusion matrix for the ResNet50 model after 40 epochs, capturing its performance on a multi-class classification task involving ten student categories. The diagonal entries, representing the number of correct predictions for each class, are predominantly high, indicating a successful classification for the majority of instances. There are minimal instances of misclassification, as shown by the very few non-zero entries off the diagonal. This matrix suggests that the ResNet50 model has achieved a high level of accuracy in distinguishing between the different student groups after the specified training period.

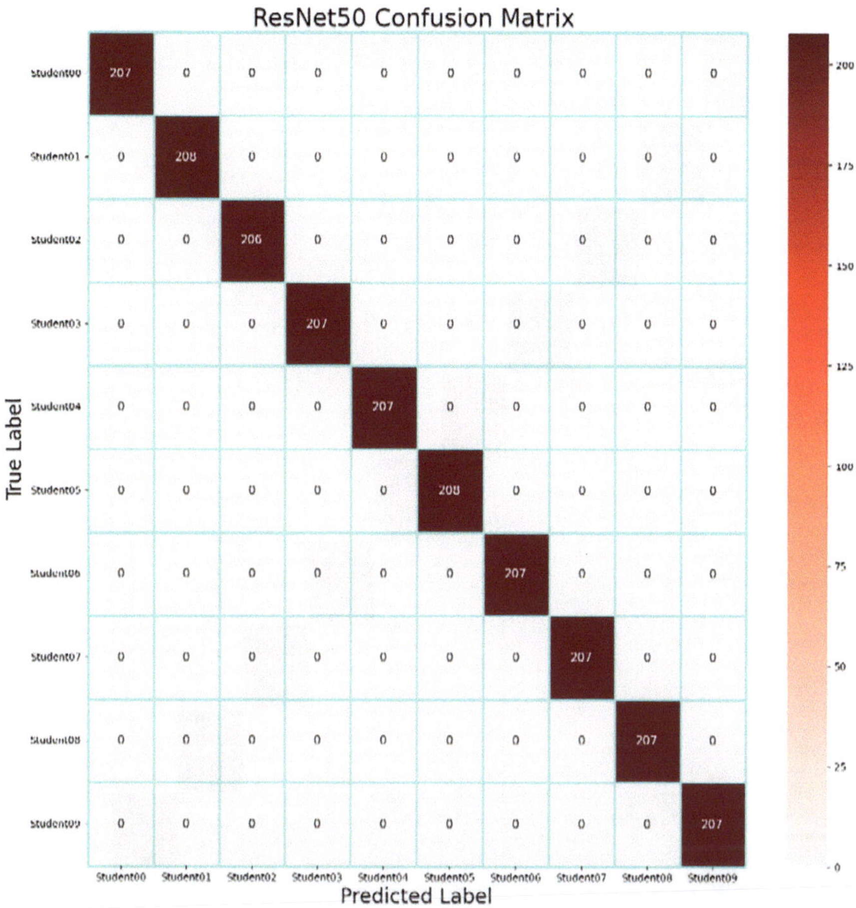

Fig. 7 Visualization of ResNet50 classification performance: confusion matrix after 30 epochs

Figure 10 presents a confusion matrix for the MobileNetV2 model after 40 epochs, with high value along the diagonal indicating accurate predictions for most student classes and scattered non-diagonal entries reflecting some misclassifications.

4.4 *Predictions of ResNet50 and MobileNetV2 Models*

Figure 11 illustrates a collection of test predictions made by the ResNet50 model after undergoing 30 training epochs. Each image represents a distinct fingerprint belonging to one of several student categories. The actual label is displayed for each sample, alongside the model's prediction and the associated confidence level. Remarkably, the model demonstrates high confidence in its predictions, with the

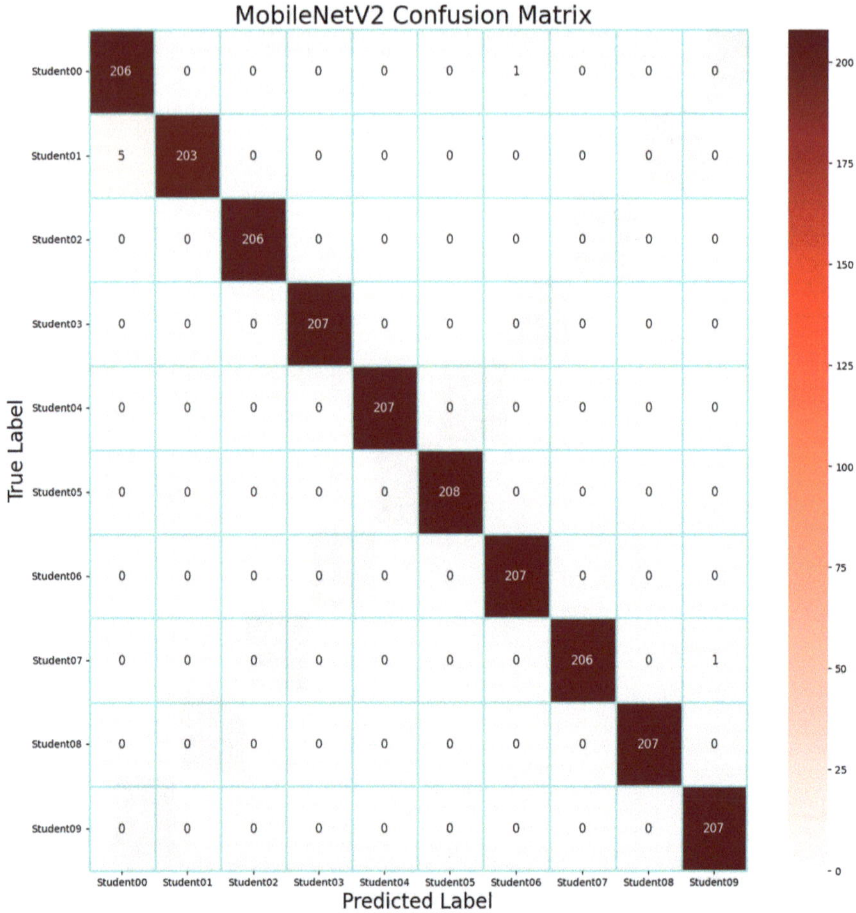

Fig. 8 Visualization of MobileNetV2 classification performance: confusion matrix after 30 epochs

probabilities consistently reaching 100%. This visual array underscores the model's accuracy in correctly identifying and classifying the fingerprints, indicative of its robust learning from the training process.

Figure 12 displays the MobileNetV2 model's test predictions after 30 epochs, showing a series of fingerprint images with corresponding actual and predicted labels. The model predicts with high confidence, generally marked at 100%, indicating precise classification across various student fingerprints.

Figure 13 showcases the ResNet50 model's fingerprint classification results after 40 training epochs. The images are labeled with their actual student category and the model's prediction, displaying a near-perfect match with high confidence levels in each prediction, indicative of the model's accuracy and reliability in this task.

Figure 14 presents test predictions of the MobileNetV2 model after 40 training epochs, illustrating fingerprint images with their actual and predicted student labels.

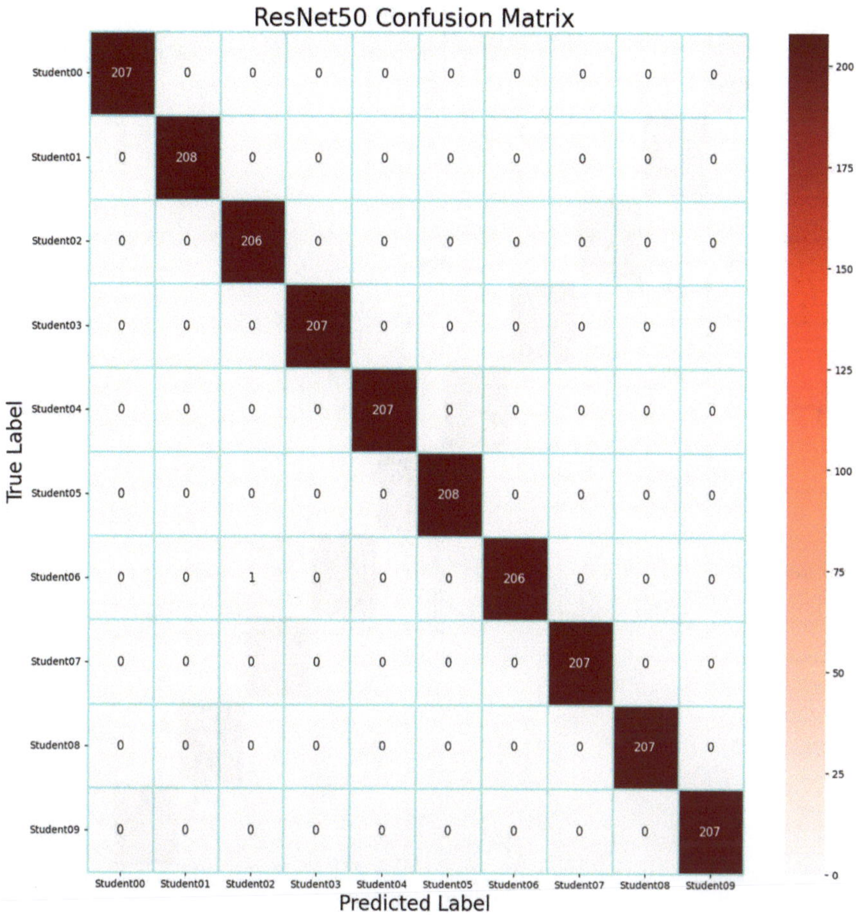

Fig. 9 Visualization of ResNet50 classification performance: confusion matrix after 40 epochs

Most predictions show 100% confidence, denoting the model's high accuracy and consistency in classification.

In the comparative analysis of the ResNet50 and MobileNetV2 models, performance metrics across two training durations—30 and 40 epochs—were examined. At 30 epochs, ResNet50 achieved perfection across all metrics, scoring 100% in accuracy, precision, recall, F1 score, and Matthews Correlation Coefficient (MCC). MobileNetV2 also performed admirably, albeit with slightly lower metrics, achieving just under 100% in the same categories.

Extending training to 40 epochs, ResNet50 maintained near-perfect scores, while MobileNetV2 displayed a notable decrease, with accuracy, precision, recall, and F1 score falling to the 98% range, and MCC to 98.13%. This decline suggests that while MobileNetV2 benefits from longer training, it may not sustain the consistency of ResNet50, which shows only a slight dip from the 30 to the 40-epoch mark.

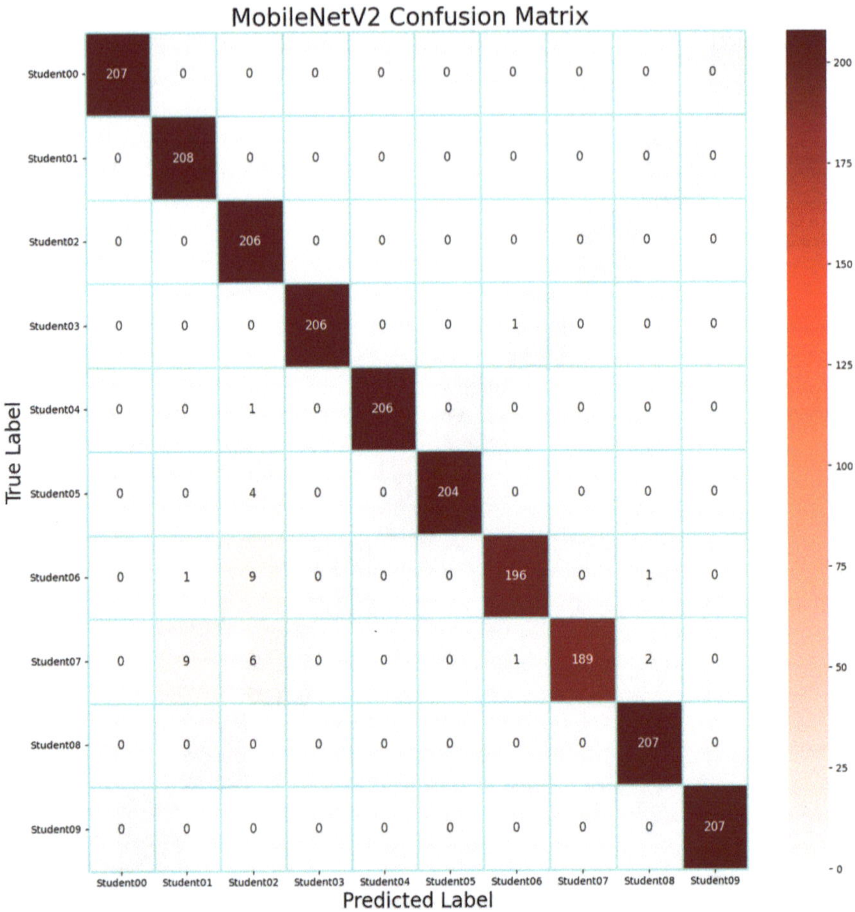

Fig. 10 Visualization of MobileNetV2 classification performance: confusion matrix after 40 epochs

The consistently high performance of ResNet50, even as the number of epochs increased, indicates robustness and a potential better fit for scenarios requiring high precision. Conversely, MobileNetV2, while still providing high performance, might be more sensitive to overfitting or other issues that could arise with more extended training. These findings highlight the importance of epoch selection in model training and suggest a nuanced approach to training duration based on the specific model architecture.

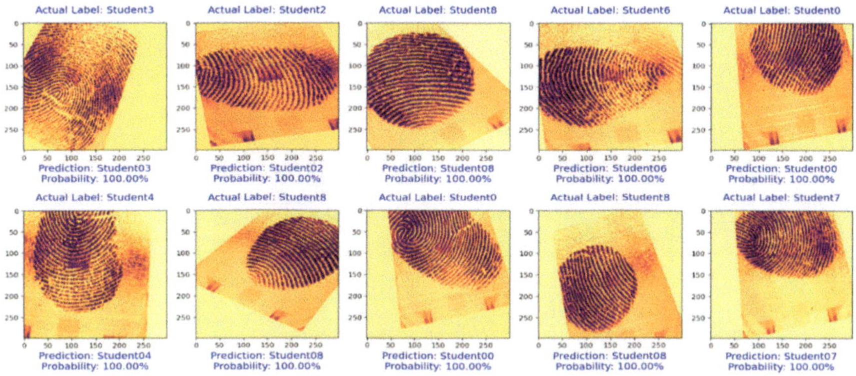

Fig. 11 Test predictions of the ResNet50 model after 30 training epochs

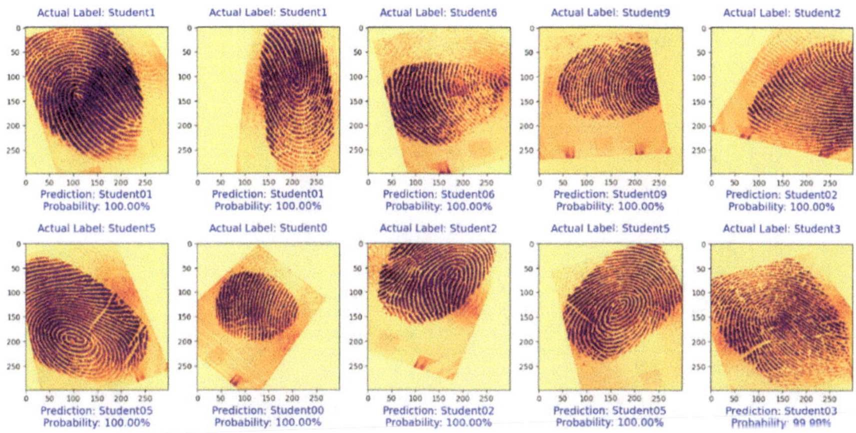

Fig. 12 Test predictions of the MobileNetV2 model after 30 training epochs

5 Conclusions

In this study, we systematically compared the performance of two deep learning models, ResNet50 and MobileNetV2, which has yielded informative insights into the efficacy of utilizing advanced neural network architectures for biometric verification in educational settings. Our comparative analysis underscores the superior proficiency of ResNet50 in fingerprint recognition tasks, achieving stellar performance metrics across the board. Its exceptional ability to maintain accuracy over increased epochs delineates its potential for deployment in student attendance systems that demand high precision and reliability.

While the MobileNetV2 model demonstrated noteworthy performance, it was slightly outshone by ResNet50, especially as the training epochs were extended. This performance differential highlights the importance of model selection based

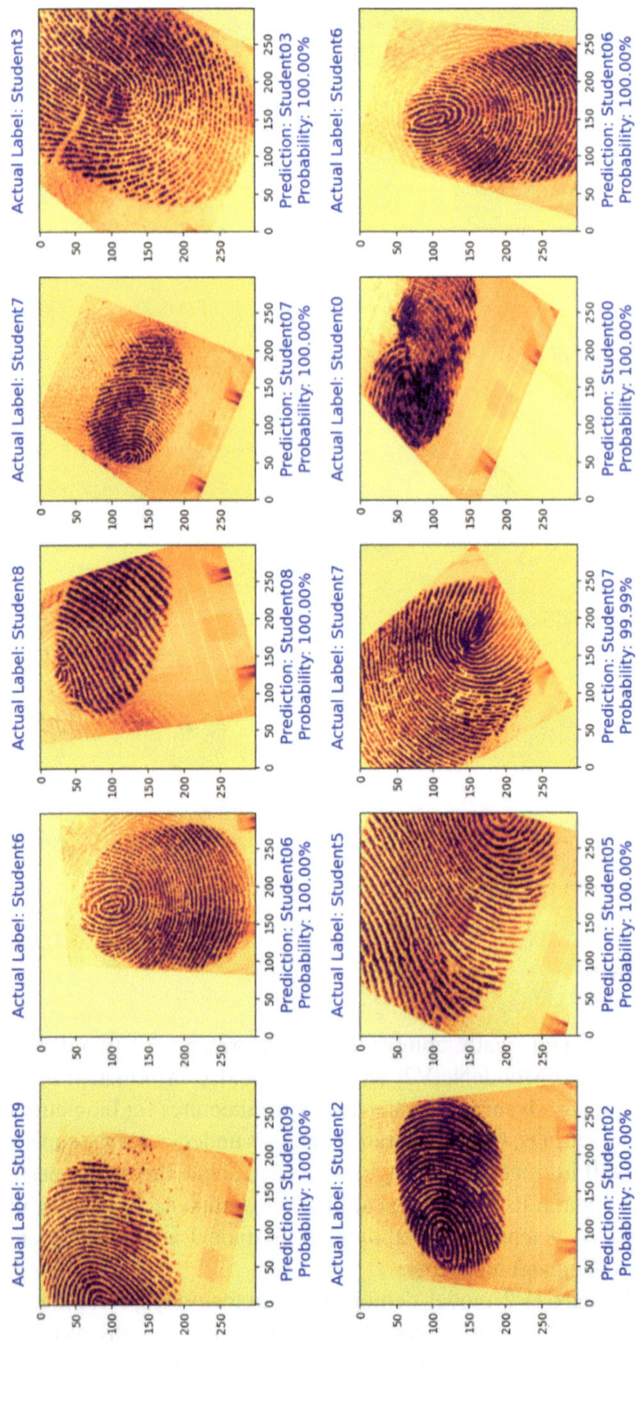

Fig. 13 Test predictions of the ResNet50 model after 40 training epochs

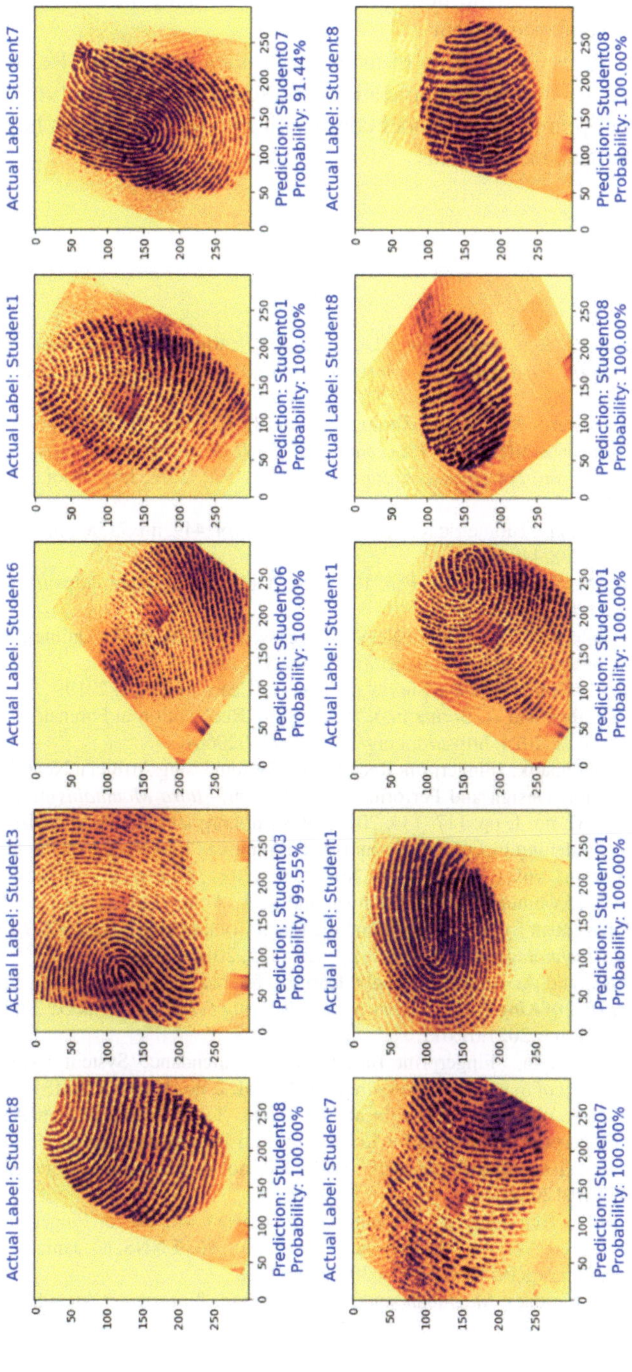

Fig. 14 Test predictions of the MobileNetV2 model after 40 training epochs

on the specific requirements of the task. The application of transfer learning in this domain effectively enhanced the models' capabilities, enabling the rapid deployment of sophisticated attendance systems.

This study confirms the viability of using pre-trained models like ResNet50 for high-accuracy tasks in real-world applications and sets a precedent for future studies to explore further optimizations, model combinations, and applications beyond the scope of student attendance systems.

References

1. K. He, X. Zhang, S. Ren, and J. Sun, "Deep residual learning for image recognition," in *Proceedings of the IEEE conference on computer vision and pattern recognition*, 2016, pp. 770–778.
2. M. Sandler, A. Howard, M. Zhu, A. Zhmoginov, and L. C. Chen, "MobileNetV2: Inverted Residuals and Linear Bottlenecks," *Proceedings of the IEEE Computer Society Conference on Computer Vision and Pattern Recognition*, pp. 4510–4520, 2018, https://doi.org/10.1109/CVPR.2018.00474.
3. A. K. Jain, "Technology: Biometric recognition," *Nature*, vol. 449, no. 7158, pp. 38–40, 2007, https://doi.org/10.1038/449038a.
4. "ResNet and ResNetV2." Accessed: Feb. 10, 2024. [Online]. Available: https://keras.io/api/applications/resnet/#resnet50-function
5. "MobileNet, MobileNetV2, and MobileNetV3." Accessed: Feb. 10, 2024. [Online]. Available: https://keras.io/api/applications/mobilenet/#mobilenetv2-function
6. F. Zhuang *et al.*, "A Comprehensive Survey on Transfer Learning," Nov. 2019.
7. N. Martins, J. S. Silva, and A. Bernardino, "Fingerprint Recognition in Forensic Scenarios," *Sensors*, vol. 24, no. 2, 2024, https://doi.org/10.3390/s24020664.
8. P. Marák and A. Hambalík, "Fingerprint Recognition System Using Artificial Neural Network as Feature Extractor: Design and Performance evaluation," *Tatra Mountains Mathematical Publications*, vol. 67, no. 1, pp. 117–134, 2016, https://doi.org/10.1515/tmmp-2016-0035.
9. O. Akinduyite, A. Adetunmbi, O. Olabode, and O. Ibidunmoye, "Fingerprint-Based Attendance Management System," vol. 1, pp. 100–105, Nov. 2013, https://doi.org/10.12691/jcsa-1-5-4.
10. A. I. Abdullateef, Ekwemuka, Itopa, Makinwa, and S. A. Alim, "Fingerprint Based Student Attendance Management System With Automatic Excel Computation," *LAUTECH Journal of Engineering and Technology*, vol. 12, no. 2, pp. 123–135, 2018.
11. R. Kute, V. Vyas, and A. Anuse, "Transfer learning for face recognition using fingerprint biometrics," *Journal of King Saud University - Engineering Sciences*, no. xxxx, 2021, https://doi.org/10.1016/j.jksues.2021.07.011.
12. P. Verma and N. Gupta, "Fingerprint Based Student Attendance System Using GSM," *International Journal of Science and Research (IJSR)*, vol. 2, no. 10, pp. 128–131, 2013.
13. N. Z. Lamin, W. N. A. W. Jusoh, J. Zainudin, and H. Samad, "Implementing Student Attendance System Using Fingerprint Biometrics for Kolej Universiti Poly-Tech Mara," *IOP Conference Series: Materials Science and Engineering*, vol. 1062, no. 1, 2021, https://doi.org/10.1088/1757-899X/1062/1/012037.
14. T. N. Do and M. T. Tran-Nguyen, "Training Deep Network Models for Fingerprint Image Classification," *Lecture Notes in Networks and Systems*, vol. 363 LNNS, no. January, pp. 327–337, 2022, https://doi.org/10.1007/978-3-030-92666-3_28.
15. M. Alhothaily, M. Alradaey, M. Oqbah, and A. El-Kustaban, "Fingerprint Attendance System for Educational Institutes," *Journal of Science and Technology*, vol. 20, no. 1, pp. 34–44, 2015, https://doi.org/10.20428/jst.20.1.4.

16. I. of A. at the C. A. of Sciences, "CASIA Fingerprint Image Database." Accessed: Feb. 01, 2024. [Online]. Available: http://biometrics.idealtest.org/findTotalDbByMode.do?mode=Fingerprint#/datasetDetail/7
17. K. He and J. Sun, "Deep Residual Learning for Image Recognition," pp. 1–9.
18. L. Zhang, Y. Bian, P. Jiang, and F. Zhang, "A Transfer Residual Neural Network Based on ResNet-50 for Detection of Steel Surface Defects," *Applied Sciences (Switzerland)*, vol. 13, no. 9, 2023, https://doi.org/10.3390/app13095260.
19. K. Dong, C. Zhou, Y. Ruan, and Y. Li, "MobileNetV2 Model for Image Classification," in *2020 2nd International Conference on Information Technology and Computer Application (ITCA)*, 2020, pp. 476–480. https://doi.org/10.1109/ITCA52113.2020.00106.
20. H. M and S. M.N, "A Review on Evaluation Metrics for Data Classification Evaluations," *International Journal of Data Mining & Knowledge Management Process*, vol. 5, no. 2, pp. 01–11, 2015, https://doi.org/10.5121/ijdkp.2015.5201.
21. Ferrer, L.: "Analysis and Comparison of Classification Metrics," pp. 1–36, 2022.

Detecting and Preventing Fraud in Banks Using Artificial Intelligence (AI): A Literature Review

Hamid Mohsin Jadah, Sakina Fatima, and Noor Hashim Mohammed Al-Husainy

Abstract Fraud detection in financial services has been a perennial challenge, requiring constant adaptation to ever-evolving landscape of financial crimes. Advances in technology, particularly artificial intelligence (AI) and machine learning (ML), offer new avenues for enhancing detection and prevention of fraud. As fraudsters become more sophisticated, the need for innovative and robust solutions becomes increasingly critical to protect financial institutions and their customers. This paper discusses a literature review that effectively convey use of AI for fraud detection and prevention in banking sector through review of existing literature. Fraud detection using AI in banking swiftly recognize irregularities, analyze patterns, and assign risk scores to customers. In today's era, fraud is pervasive across various sectors, particularly in financial services. The rise of technologies like AI and big data is revolutionizing fraud detection by leveraging ML to analyze vast amounts of data and identify suspicious activities. Fraudulent activities pose significant threat to various sectors of society, including finance, e-commerce, healthcare, and more. Traditional methods of fraud detection and prevention often fall short in identifying and mitigating sophisticated and rapidly evolving fraud schemes.

Keywords Literature review · Fraud detection · Financial services · Financial crimes · Artificial intelligence (AI) · Banking sector

H. M. Jadah (✉)
Department of Finance and Banking, College of Administration and Economics, University of Kerbala, Karbala, Iraq
e-mail: hamed.m@uokerbala.edu.iq

S. Fatima
Department of Computer Science, All Saint's College of Technology, Rajiv Gandhi Proudyogiki Vishwavidyalaya, Bhopal, Madhya Pradesh, India
e-mail: sakina.fatima.imam@gmail.com

N. H. M. Al-Husainy
Babylon Departments, Imam Al-Kadhum College for Islamic Studies, Babylon, Iraq
e-mail: nourhashim@iku.edu.iq

M. Azrour and A. Alabdulatif (eds.), *AI-Driven Security for Next-Generation IoT Systems*, https://doi.org/10.1007/978-3-032-08784-3_14

1 Introduction

With the rise of technology and ongoing economic growth in modern society, fraud has become increasingly prevalent in the financial industry, costing institutions and consumers hundreds of billions of dollars annually. Fraudsters continuously evolve their methods to exploit vulnerabilities in existing prevention measures, particularly targeting financial sector. These crimes include credit card fraud, healthcare and automobile insurance fraud, money laundering, securities and commodities fraud, and insider trading. Fraud prevention systems alone do not provide sufficient security against these criminal activities [1–3]. Consequently, necessity for fraud detection systems that can identify fraudulent acts after they have occurred—and potential cost savings from such detection—has never been clearer. Researchers have intensively studied anomaly detection techniques for this purpose over last few decades, employing statistical, artificial intelligence, and ML models. In recent years, development of AI for detecting fraudulent banking operations has garnered significant attention. This focus is driven by increasing number of fraudulent activities in banking industry, leading to substantial financial losses for both banks and their customers. AI-based systems offer potential to identify and prevent fraudulent activities in real time, providing considerable advantage over traditional fraud detection methods. As banking industry expands, it introduces wealth of new financial products that simplify transactions, enabling customers to send and receive money effortlessly, apply for loans, and even access funds from home [4–7]. However, this convenience also makes banks more vulnerable to scammers. AI excels at swiftly sifting through vast datasets to identify unusual or suspicious patterns that may indicate fraud. To identify suspicious behaviors, AI-powered fraud detection systems analyze massive datasets using variety of ML models to detect trends and anomalies. These algorithms continuously learn from new data, enhancing their ability to identify and prevent fraudulent activities. By proactively implementing these systems, organizations can safeguard integrity and security of their transactions.

This paper provides comprehensive literature review on application of AI in detecting and preventing fraud in banking sector. It discusses growing necessity for advanced fraud detection systems due to increasing sophistication of fraudulent activities. Review focuses on various AI techniques, their effectiveness, implementation challenges, and future research directions. Objective is to provide an insightful analysis of how AI can enhance fraud detection and prevention, thereby safeguarding financial institutions and their customers. Fraudulent activities in banking sector have been persistent challenge, leading to significant financial losses for banks and their customers. Traditional fraud detection methods, which often rely on static rule-based systems and manual reviews, are increasingly proving inadequate against sophisticated and rapidly evolving fraud schemes. This has led to growing interest in leveraging AI to enhance fraud detection and prevention capabilities. This paper reviews existing literature on AI-based fraud detection in banking, exploring effectiveness of various AI techniques and identifying areas for future research.

2 Literature Survey

Advent of digital banking and proliferation of online financial transactions have heightened need for sophisticated fraud detection and prevention mechanisms in banks. AI has emerged as pivotal tool in combating financial fraud [8–12]. AI has significantly advanced capabilities of fraud detection and prevention in banks. ML, DL, and hybrid models have shown great potential in identifying and mitigating fraudulent activities. However, challenges like data quality, adversarial attacks, and ethical concerns must be addressed to fully realize benefits of AI in this domain. Continued research and innovation are essential to stay ahead of evolving fraud tactics and ensure security and integrity of financial systems. Solomiia Fedushko et al. [13] anticipated model, which is based on an artificial neural network, effectively advances accuracy of fraudulent transaction detection. Outcomes of dissimilar algorithms are visualized, and logistic regression algorithm executes superlative, with an output AUC value of approximately 0.946. Stacked generalization depicts better AUC of 0.954. Recognition of banking fraud using AI algorithms is topical matter in our digital society. They focused on applying ML models to identify fraudulent banking transactions. They also applied preprocessing procedures to compare and select most effective consequences from bank data. Bao et al. [14] used framework (data, process, and evaluation criterion) to review some of practical considerations that may affect enactment of ML models to predict fraud. Farman et al. [15] explores role of AI in revolutionizing fraud detection and prevention. They delve into key AI techniques, their applications in different industries, challenges, and future of AI-driven fraud prevention systems. ANN has revolutionized fraud detection and prevention across various industries. Techniques like ML, DL, and natural language processing enable organizations to identify and mitigate fraud more effectively than ever. However, challenges like data privacy, adversarial attacks, scalability, and human–machine collaboration remain. As technology advances, AI's potential to combat fraud will grow even further. Key trends shaping future of fraud prevention include explainable AI, enhanced biometric authentication, blockchain technology, real-time data processing, and cross-industry collaboration. As AI continues to evolve, it will play crucial role in safeguarding integrity of financial systems, e-commerce platforms, healthcare data, and digital assets. Kamuangu [16] their study thoroughly explores advanced approaches for addressing financial fraud, focusing on effectiveness of ML and AL. By recognizing shortcomings of outdated methods, examination aims to analyze current landscape, closely examining efficiency and limitations of ML and AI techniques while outlining intricate directions for future research. Study delves into complex history of financial fraud, revealing inherent constraints of conventional rule-based and manual detection approaches. It then highlights ML and AI, showcasing significant research and successful implementations that have revolutionized fraud detection.

3 Background and Motivation

Expansion of banking industry has introduced plethora of new financial products and services that streamline transactions. However, this convenience also increases sector's vulnerability to fraud. Traditional methods of fraud detection, like manual reviews and rule-based systems, are not only time-consuming but also prone to errors. These methods struggle to keep pace with evolving tactics of fraudsters, necessitating adoption of more advanced, automated solutions. AI, with its ability to process vast amounts of data and detect complex patterns, offers promising alternative. Banking sector combats various fraud types, including identity theft, credit card fraud, loan fraud, phishing, money laundering, and payment fraud. These categories reflect broad spectrum of tactics used by fraudsters, requiring banks to employ diverse and sophisticated countermeasures. In banking industry, AI-based fraud detection offers numerous advantages: it provides high accuracy, adapts to emerging threats, is more cost-effective than traditional methods, improves customer experience, manages risk effectively, and assists with regulatory compliance. AI is revolutionizing banking industry by enhancing efficiency, security, and customer satisfaction. As technology continues to evolve, AI's role in banking is expected to expand, offering even more innovative solutions to meet challenges and opportunities in financial sector. AI prevents fraud by analyzing data in real time to detect unusual patterns, using ML to adapt to new fraud tactics. Bank uses ML to analyze historical and real-time transaction data, identifying and flagging anomalies as potential fraud. This approach minimizes false positives and evolves with emerging fraud methods, making fraud detection systems more accurate and responsive over time.

4 Fraud Management in Banking: Detection, Prevention

With rise of technology and ongoing economic growth in modern society, fraud has become increasingly prevalent in financial industry, costing institutions and consumers hundreds of billions of dollars annually. Fraudsters continuously evolve their methods to exploit vulnerabilities in existing prevention measures, particularly targeting financial sector [17–20]. These crimes include credit card fraud, healthcare and automobile insurance fraud, money laundering, securities and commodities fraud, and insider trading. Fraud prevention systems alone do not provide sufficient security against these criminal activities. Consequently, necessity for fraud detection systems that can identify fraudulent acts after they have occurred—and potential cost savings from such detection—has never been clearer. Researchers have intensively studied anomaly detection techniques for this purpose over last few decades, employing statistical, AI, and ML models (Fig. 1).

Fraud in financial sector has surged with technological advancements and economic growth, costing institutions and consumers hundreds of billions of dollars

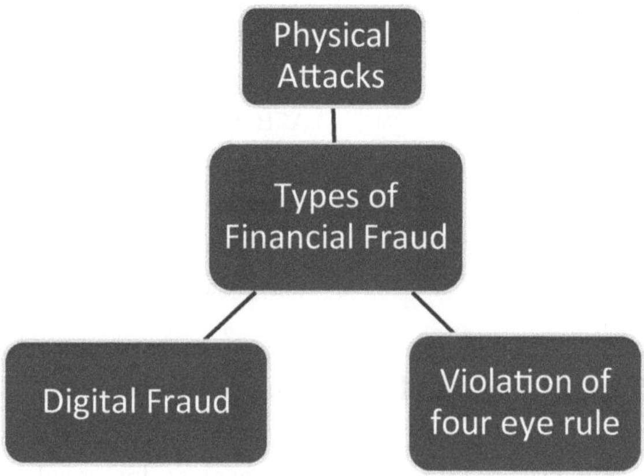

Fig. 1 Types of financial fraud

annually. This rise in fraud includes various crimes like credit card fraud, healthcare and automobile insurance fraud, money laundering, securities and commodities fraud, and insider trading. Fraudsters are continuously evolving their methods to bypass existing prevention measures.

- **Detection**: Detection involves identifying fraudulent activities after they have occurred. Anomaly detection techniques, which include statistical, AI, and ML models, are essential in identifying suspicious activities that deviate from normal patterns. Supervised learning algorithms have been popular for detecting fraud, as they can learn from labeled datasets to recognize and flag fraudulent transactions.
- **Prevention**: Prevention focuses on stopping fraud before it happens. This includes implementing robust security measures like multi-factor authentication, encryption, and real-time monitoring systems. Institutions also need to regularly update their fraud prevention strategies to address emerging threats and vulnerabilities. Educating consumers about common fraud schemes and encouraging them to report suspicious activities are also crucial steps in preventing fraud.

Therefore, effective fraud management in banking requires dual approach of detection and prevention. While detection systems help identify and mitigate impact of fraud after it occurs, prevention strategies aim to block fraudulent activities before they can cause harm. By combining advanced technologies with proactive measures, financial institutions can better protect themselves and their customers from growing threat of fraud. Table 1 mentions other types of frauds that occurs in financial industry (banks).

Table 1 Classification of financial frauds

Classification of financial frauds	
Identity theft	Payment fraud
ACH fraud	Account takeover fraud
Advanced fee fraud	Credit card fraud
Investment fraud	Consumer fraud
Fraudulent charities	Return fraud
Chargeback fraud	Cyber crime

5 Utilization of AI in Banking Sector

Banks use machine learning to analyze historical and real-time transaction data, identifying and flagging anomalies as potential fraud. This approach minimizes false positives and evolves with emerging fraud methods, making fraud detection systems more accurate and responsive over time. Fraud detection in financial services has been perennial challenge, requiring constant adaptation to ever-evolving landscape of financial crimes. Advances in technology, offers new avenues for enhancing detection and prevention of fraud. As fraudsters become more sophisticated, need for innovative and robust solutions becomes increasingly critical to protect financial institutions and their customers. AI-based systems can help banks adhere to regulatory requirements by providing accurate and timely detection of fraudulent activities, generating comprehensive audit trails, and ensuring data privacy and security through advanced protection measures. Table 2 depicts this challenge arises from several factors:

To address these challenges, financial institutions are increasingly turning to AI and ML for fraud detection. These technologies offer several benefits:

1. **Pattern Recognition**: AI and ML can analyze vast amounts of transaction data to identify patterns and anomalies that may indicate fraudulent activity. Unlike traditional rule-based systems, AI can adapt to new patterns without explicit programming.
2. **Real-Time Analysis**: AI systems can process data in real time, allowing institutions to detect and prevent fraud as it occurs. This immediate response is crucial in minimizing financial losses and protecting customers.
3. **Behavioral Analysis**: AI can build profiles of normal behavior for individual customers and flag deviations from these patterns. This helps in identifying fraud that might not trigger traditional rules.
4. **Enhanced Accuracy**: ML algorithms can continuously improve their accuracy by learning from new data. This reduces false positives (legitimate transactions flagged as fraud) and false negatives (fraudulent transactions that go undetected).
5. **Fraud Prediction**: Predictive analytics can forecast potential fraud risks based on historical data and emerging trends, enabling proactive measures.
6. **Integration with Other Technologies**: AI systems can integrate with other security measures, like biometric verification and blockchain, to create multi-layered defense against fraud.

Table 2 Arise of challenges from various factors

Arise of challenges from various factors	
Sophistication of fraudsters	Criminals continually develop new techniques to exploit vulnerabilities in financial systems. This includes phishing, identity theft, money laundering, and more advanced methods like synthetic identity fraud and cyber-attacks
Technological advancements	As technology advances, so do tools and methods available to fraudsters. Use of AI and ML by criminals to create more convincing fake identities or to automate large-scale fraud attempts increases complexity of detection
Volume of transactions	Financial institutions handle millions of transactions daily. Sheer volume makes it difficult to manually review each transaction for signs of fraud. Automated systems need to be both highly accurate and efficient to handle this load
Regulatory changes	Financial regulations frequently change, requiring institutions to adapt their fraud detection systems to remain compliant. This includes implementing measures to detect fraud in real time while ensuring that legitimate transactions are not unduly delayed
Diverse fraud tactics	Fraud can occur in numerous ways, including card-not-present transactions, account takeovers, insider fraud, and wire fraud. Each type of fraud requires different detection and prevention strategies
Globalization	With globalization of financial services, transactions cross borders, making it challenging to track and prevent fraud due to differing regulations, standards, and security practices across countries

Hence, dynamic and evolving nature of financial crimes necessitates continuous improvement and adaptation in fraud detection strategies. Integration of AI and ML provides powerful toolset for financial institutions to stay ahead of fraudsters, ensuring security and integrity of financial services.

6 Role of AI in Transforming Banking Sector

Banks implement fraud prevention through encryption, two-factor authentication, AI-driven anomaly detection, and real-time monitoring. They also conduct regular security audits, educate employees and customers on safe practices, and collaborate with industry partners to share intelligence on emerging threats. Banking sector combats various fraud types, including identity theft, credit card fraud, loan fraud, phishing, money laundering, and payment fraud. These categories reflect broad spectrum of tactics used by fraudsters, requiring banks to employ diverse and sophisticated countermeasures. AI is transforming banking sector in numerous ways, driving efficiency, enhancing customer experience, and improving security. Integration of AI and ML in fraud detection is transforming financial industry.

Role of AI in Banking	
Fraud detection and prevention	Regulatory compliance
Customer service	Investment and trading
Credit scoring and risk management	Fraud detection in credit card transaction
Personalized banking	Loan approval automation
Operational efficiency	Sentiment analysis
Predictive analytics	

1. **Fraud Detection and Prevention**: AI and ML algorithms analyze large volumes of transaction data to identify patterns and anomalies that may indicate fraudulent activity. These systems can process data in real time, allowing banks to detect and respond to fraud instantly. ML models improve over time, reducing false positives and increasing accuracy.
2. **Customer Service**: AI-powered chatbots and virtual assistants provide 24/7 customer service, handling routine inquiries, assisting with transactions, and providing financial advice. These AI systems enhance customer experience by offering quick, personalized responses and freeing up human agents to handle more complex issues.
3. **Credit Scoring and Risk Management**: AI models analyze wide range of data sources, including credit history, transaction behavior, and social media activity, to assess creditworthiness more accurately. This allows banks to make better lending decisions, reduce default rates, and offer personalized loan terms.
4. **Personalized Banking**: AI analyzes customer data to offer personalized financial products and services. This includes personalized investment advice, tailored savings plans, and customized offers that align with individual financial goals and behaviors.
5. **Operational Efficiency**: AI automates routine administrative tasks like data entry, document verification, and compliance reporting. This reduces operational costs, minimizes errors, and allows staff to focus on higher-value activities.
6. **Predictive Analytics**: AI and ML models predict customer needs and behaviors by analyzing historical data. For instance, banks can forecast customer attrition and take proactive measures to retain clients or anticipate cash flow needs to manage liquidity more effectively.
7. **Regulatory Compliance**: AI systems help banks comply with complex regulatory requirements by automating monitoring and reporting of transactions. This includes anti-money laundering (AML) checks, Know Your Customer (KYC) processes, and other regulatory reporting tasks.
8. **Investment and Trading**: AI algorithms analyze market data and trends to make trading decisions and optimize investment portfolios. These systems can process information much faster than humans, identifying opportunities and risks in real time.

9. **Fraud Detection in Credit Card Transactions**: AI systems monitor credit card transactions to detect unusual spending patterns that may indicate fraud. These systems can flag suspicious transactions and alert customers or block transactions until further verification.
10. **Loan Approval Automation**: AI speeds up loan approval process by automating evaluation of loan applications. This involves assessing credit risk, verifying documents, and making approval decisions based on predefined criteria and predictive analytics.
11. **Sentiment Analysis**: AI analyzes customer feedback, social media activity, and other sources to gauge customer sentiment. This helps banks understand customer satisfaction levels, identify areas for improvement, and tailor their services accordingly.

AI-based fraud detection systems offer significant advantages over traditional methods, making them more efficient in combating fraud. These systems leverage ML algorithms that can continuously learn from new data and adapt to evolving fraud patterns, ensuring they stay ahead of fraudsters' tactics. By analyzing vast volumes of data and identifying intricate patterns, AI can improve accuracy, reducing false positives and negatives. Moreover, AI systems can process transactions in real time, enabling faster fraud detection and prevention. Their ability to handle massive amounts of data and transactions makes them highly scalable, and suitable for large-scale operations (Table 3).

Moreover, AI is revolutionizing banking industry by enhancing efficiency, security, and customer satisfaction. As technology continues to evolve, AI's role in banking is expected to expand, offering even more innovative solutions to meet challenges and opportunities in financial sector. AI prevents fraud by analyzing data in real time to detect unusual patterns, using ML to adapt to new fraud tactics. It automates detection, reduces manual review needs, and enhances response speed to threats, thereby effectively safeguarding assets and customer trust (Table 4).

Table 3 Benefits of AI in banking

Enhanced security	Improved fraud detection and prevention capabilities
Cost savings	Automation of routine tasks reduces operational costs
Improved customer experience	Personalized services and faster response times
Better decision-making	Data-driven insights lead to more informed decisions
Regulatory compliance	Automated compliance reduces risk of regulatory breaches

Table 4 Challenges and considerations

Data privacy and security	Ensuring protection of sensitive customer data
Bias and fairness	Addressing potential biases in AI algorithms to ensure fair treatment
Integration	Integrating AI systems with existing banking infrastructure
Regulatory compliance	Navigating regulatory landscape for AI applications in banking

7 Implementation Challenges

Despite potential of AI in fraud detection, several challenges must be addressed to ensure effective implementation (Fig. 2):

- **Data Privacy and Security**: Ensuring privacy and security of sensitive financial data is paramount. AI systems must comply with regulatory requirements and implement robust data protection measures.
- **Adversarial Attacks**: Fraudsters may attempt to deceive AI systems by manipulating data inputs. Developing resilient models that can withstand adversarial attacks is crucial.
- **Scalability**: AI systems must be scalable to handle the growing volume of transactions in banking sector.
- **Human–Machine Collaboration**: Effective fraud detection often requires a combination of automated systems and human expertise. Balancing this collaboration is essential for optimal results.

Evaluation Metrics: To assess the performance of AI-based fraud detection systems, various metrics are used to enhance in evaluating effectiveness (Fig. 3).

- **Accuracy**: Proportion of correctly identified fraud cases.
- **Precision**: Proportion of true fraud cases among detected cases.
- **Recall**: Proportion of detected fraud cases among actual fraud cases.
- **F1 Score**: Harmonic mean of precision and recall, providing balance between two.

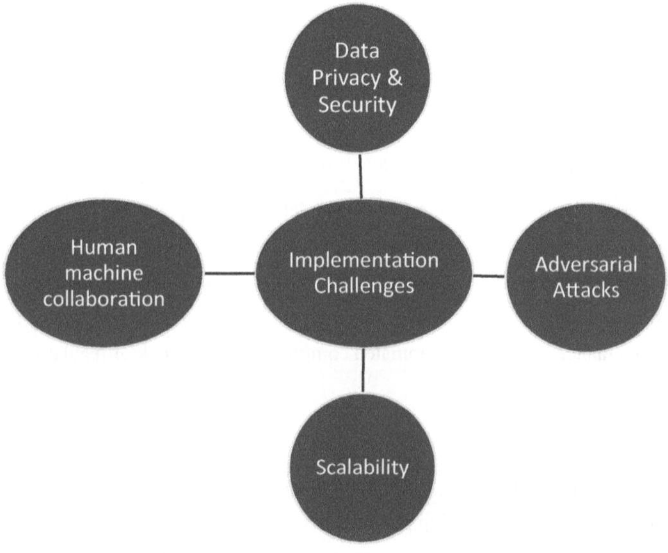

Fig. 2 Implementation challenges that must be addressed to ensure effective implementation

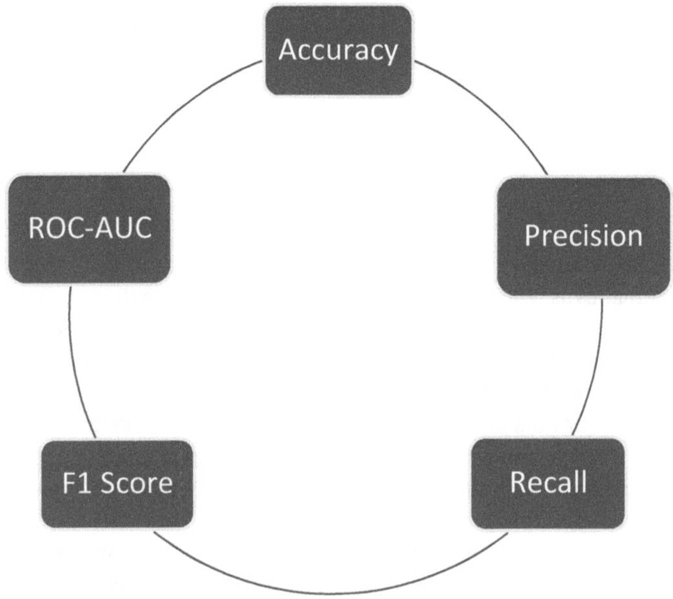

Fig. 3 Effective metrics to evaluate effectiveness of AI models

- **ROC-AUC**: Area under receiver operating characteristic curve, indicating model's ability to distinguish between legitimate and fraudulent transactions.

8 Future Research Directions

AI has transformed fraud detection and prevention in banking sector by providing more accurate, efficient, and adaptive solutions. While challenges related to data privacy, adversarial attacks, scalability, and human–machine collaboration persist, ongoing advancements in AI technology promise to further enhance fraud detection capabilities. By leveraging AI, banks can better protect their assets and customers from ever-evolving threat of fraud. Future of AI in fraud detection holds several promising directions:

- **Explainable AI (XAI)**: Developing AI systems that provide transparent and interpretable results to enhance trust and compliance.
- **Enhanced Biometric Authentication**: Integrating advanced biometric authentication methods, such as facial recognition and fingerprint scanning, with AI-based fraud detection.
- **Blockchain Technology**: Leveraging blockchain's immutability and transparency to enhance security and traceability of transactions.
- **Real-Time Data Processing**: Improving AI systems' ability to process and analyze transaction data in real time for immediate fraud detection.

- **Cross-Industry Collaboration**: Fostering collaboration between financial institutions, technology providers, and regulatory bodies to develop comprehensive fraud detection frameworks.

9 AI Techniques in Fraud Detection

Banks uses ML to analyze historical and real-time transaction data, identifying and flagging anomalies as potential fraud. They minimize false positives and evolves with emerging fraud methods, making fraud detection systems more accurate and responsive over time. Fraud detection using AI in banking swiftly recognize irregularities, analyze and assign risk scores to customers. Integration of AI and ML provides powerful toolset for financial institutions to stay ahead of fraudsters, ensuring security and integrity of financial services. By leveraging these AI techniques, banks can significantly enhance their ability to detect and prevent fraud, thereby safeguarding their assets and maintaining customer trust (Fig. 4).

These techniques collectively enhance ability to detect and prevent fraudulent activities by providing more accurate, adaptive, and comprehensive analysis compared to traditional methods. To address these challenges, financial institutions are increasingly turning to AI and ML for fraud detection. These technologies offer several benefits:

1. **Pattern Recognition**: AI and ML can analyze vast amounts of transaction data to identify patterns and anomalies that may indicate fraudulent activity. Unlike traditional rule-based systems, AI can adapt to new patterns without explicit programming.
2. **Real-Time Analysis**: AI systems can process data in real time, allowing institutions to detect and prevent fraud as it occurs. This immediate response is crucial in minimizing financial losses and protecting customers.

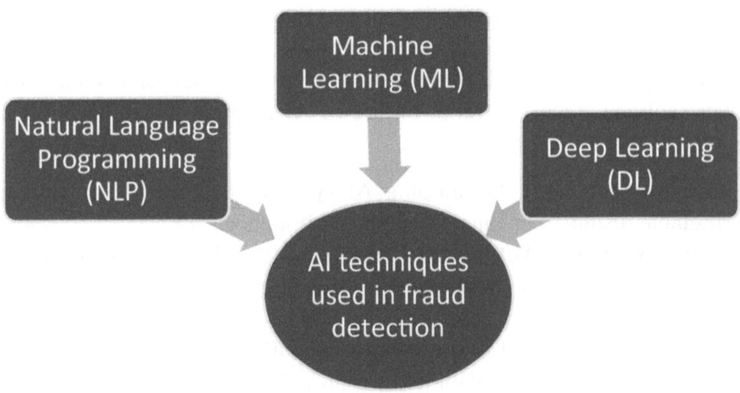

Fig. 4 Common AI techniques used in fraud detection

3. **Behavioral Analysis**: AI can build profiles of normal behavior for individual customers and flag deviations from these patterns. This helps in identifying fraud that might not trigger traditional rules.
4. **Enhanced Accuracy**: ML algorithms can continuously improve their accuracy by learning from new data. This reduces false positives (legitimate transactions flagged as fraud) and false negatives (fraudulent transactions that go undetected).
5. **Fraud Prediction**: Predictive analytics can forecast potential fraud risks based on historical data and emerging trends, enabling proactive measures.
6. **Integration with Other Technologies**: AI systems can integrate with other security measures, such as biometric verification and blockchain, to create multi-layered defense against fraud.

Machine Learning (ML): ML involves training algorithms on historical data to identify patterns indicative of fraud. Techniques such as supervised learning, unsupervised learning, and reinforcement learning are widely used in fraud detection. Supervised learning algorithms, such as decision trees, support vector machines (SVM), and neural networks, are trained on labeled datasets to distinguish between legitimate and fraudulent transactions. Unsupervised learning techniques, such as clustering and anomaly detection, identify outliers in data that may indicate fraudulent activity.

Deep Learning: DL, a more advanced subset of ML, employs neural networks with multiple layers to analyze complex datasets. Convolutional neural networks (CNNs) and recurrent neural networks (RNNs) are particularly effective in detecting fraud due to their ability to capture intricate patterns and temporal dependencies in transaction data.

Natural Language Processing (NLP): Natural language processing (NLP) is used to analyze textual data, such as transaction descriptions and customer communications, to detect potential fraud. NLP techniques can identify suspicious keywords, sentiment, and context, enhancing overall fraud detection process.

10 Conclusion

This comprehensive literature review underscores critical importance of applying AI in detecting and preventing fraud in banking sector. Growing sophistication and frequency of fraudulent activities necessitate advanced fraud detection systems. Various AI techniques, such as machine learning, neural networks, and anomaly detection, have demonstrated significant potential in enhancing effectiveness of fraud detection and prevention. Despite promise, implementing AI-based fraud detection systems is fraught with challenges, including data privacy concerns, integration issues with existing systems, and need for continuous learning and adaptation. However, benefits of leveraging AI to safeguard financial institutions and their customers far outweigh these challenges, as AI enhances accuracy and response times in detecting fraudulent activities.

Future research should focus on refining AI models to improve accuracy, addressing ethical and legal implications, and developing scalable solutions that can adapt to evolving fraud patterns. Strategic implementation of AI, supported by human expertise and traditional methods, is essential for creating robust fraud detection framework.

Ultimately, collaborative effort among banks, AI researchers, and technology providers is crucial to developing innovative solutions that effectively combat financial fraud. By harnessing ongoing advancements in AI technology and data analytics, financial institutions can continuously improve their fraud detection capabilities, thereby maintaining trust and confidence among their customers.

References

1. Lucía Rey-Ares, Sara Fernández-López, Marcos Álvarez-Espiño, The role of financial literacy in consumer financial fraud exposure (via email) and victimisation: evidence from Spain, International Journal of Bank Marketing, https://doi.org/10.1108/IJBM-03-2023-0169, (2024).
2. Martynas Damulis, Investigations of Financial Fraud, Theory and Practice of Illegitimate Finance, https://doi.org/10.4018/979-8-3693-1190-5.ch011, (203–221), (2023).
3. Vivekrabinson K, Bharath Singh J, Rajesh Kumar S, Vijay M, Vijayakumar D, Shanmuga Raja B, Blockchain Enabled Real Estate Property Transactions using NFT: An Approach, 2023 International Conference on Research Methodologies in Knowledge Management, Artificial Intelligence and Telecommunication Engineering (RMKMATE), https://doi.org/10.1109/RMKMATE59243.2023.10369785, (1–6), (2023).
4. Bar Lev E, Maha L-G and Topliceanu S-C (2022) Financial frauds' victim profiles in developing countries. *Front. Psychol.* 13:999053. https://doi.org/10.3389/fpsyg.2022.999053
5. Kulal, Abhinandan and Bhat, Swathi, Online Scams in Financial Market-A study with reference to Indian Financial Market (October 21, 2022).
6. Ali A, Abd Razak S, Othman SH, Eisa TAE, Al-Dhaqm A, Nasser M, Elhassan T, Elshafie H, Saif A. Financial Fraud Detection Based on Machine Learning: A Systematic Literature Review. *Applied Sciences*. 2022; 12(19):9637.
7. Karpoff JM. The future of financial fraud. Journal of Corporate Finance. 2021 Feb;66:101694.
8. Sina Ahmadi. Open AI and its Impact on Fraud Detection in Financial Industry. Journal of Knowledge Learning and Science Technology ISSN: 2959–6386 (online), 2023, 2 (3), pp.263–281.
9. Reurink, Arjan. (2016). Financial Fraud: A Literature Review. MPIfG Discussion Paper Series.
10. Afjal M, Salamzadeh A, Dana L-P. Financial Fraud and Credit Risk: Illicit Practices and Their Impact on Banking Stability. *Journal of Risk and Financial Management*. 2023; 16(9):386.
11. Kot, M. (2024). Financial frauds throughout the years – literature review. Research Papers in Economics and Finance, 8(1), 39–60.
12. Ahmad Ahed Bader, Yousef A. Abu Hajar, Sulaiman Raji Sulaiman Weshah, Bisan Khalil Almasri, Predicting Risk of and Motives behind Fraud in Financial Statements of Jordanian Industrial Firms Using Hexagon Theory, Journal of Risk and Financial Management, https://doi.org/10.3390/jrfm17030120, **17**, 3, (120), (2024).
13. Mytnyk B, Tkachyk O, Shakhovska N, Fedushko S, Syerov Y. Application of Artificial Intelligence for Fraudulent Banking Operations Recognition. *Big Data and Cognitive Computing*. 2023; 7(2):93.
14. Bao, Yang and Hilary, Gilles and Ke, Bin, Artificial Intelligence and Fraud Detection (November 24, 2020). Babich V, Birge J, Hilary G (eds) Innovative Technology at the interface of Finance and Operations. Springer Series in Supply Chain Management, forthcoming, Springer Nature.

15. Farman, Muhammad and Abbas, Asad. (2023). Artificial Intelligence for fraud detection and prevention.
16. Kamuangu, Paulin and K K, Paul. (2024). A Review on Financial Fraud Detection using AI and Machine Learning. Journal of Economics Finance and Accounting Studies. 6. 67–77. https://doi.org/10.32996/jefas.6.1.7.
17. Kadoya Y, Khan MSR, Narumoto J and Watanabe S (2021) Who Is Next? A Study on Victims of Financial Fraud in Japan. *Front. Psychol.* 12:649565. https://doi.org/10.3389/fpsyg.2021.649565
18. Rathinaraj, Daniel and Chendroyaperumal, Chendrayan, Financial Fraud, Cyber Scams and India – A Small Survey of Popular Recent Cases (May 12, 2010).
19. Zhu X, Ao X, Qin Z, Chang Y, Liu Y, He Q, Li J. Intelligent financial fraud detection practices in post-pandemic era. Innovation (Camb). 2021 Oct 20;2(4):100176. https://doi.org/10.1016/j.xinn.2021.100176. PMID: 34806059; PMCID: PMC8581570.
20. Kot, M. (2024). Financial frauds throughout the years – literature review. *Research Papers in Economics and Finance*, *8*(1).

AI-Driven Digital Forensics in the Next-Generation Internet of Medical Things (IoMT)

Muhammad Ahsan Qureshi and Maria Zuraiz

Abstract The Internet of Medical Things (IoMT) is revolutionizing healthcare systems by creating a network of IoT devices and infrastructures that enable not only continuous patient monitoring but also decision-making based on the data collected. However, this makes the security perimeter larger, which in turn, increase the risk of cyberattacks such as unauthorized access to data or data manipulation. This chapter highlights that IoMT security is a crucial aspect of the overall cybersecurity in healthcare systems and outlines the principal cybersecurity and AI-driven digital forensics challenges they have encountered. In addition to that, it also describes evolving trends including use cases and methodologies for AI-driven IoMT forensics, and future directions. Finally, the chapter gives an overview of global health standards for IoMT security compliance, and points to the fact that there is a need for skilled experts and sustainable technologies to protect future digital healthcare.

Keywords Cybersecurity · Digital forensics · Internet of Things · Smart healthcare · Artificial intelligence · Machine learning · Blockchain · Data privacy

1 Introduction

Internet of Medical Things (IoMT) has changed the face of the healthcare sector with end-to-end connectivity among patients, healthcare professionals, and medical equipment. From implants in the heart to wearable health monitors, IoMT is the backbone of modern healthcare networks, increasing accuracy and lowering the

M. A. Qureshi (✉)
Department of Computer Science, Faculty of Computing and Artificial Intelligence, Air University Islamabad, Aerospace and Aviation Campus, Kamra, Pakistan
e-mail: ahsan@aack.au.edu.pk

M. Zuraiz
Department of Robotics and Artificial Intelligence, SZABIST University, Islamabad, Pakistan
e-mail: maria.zuraiz@szabist-isb.edu.pk

M. Azrour and A. Alabdulatif (eds.), *AI-Driven Security for Next-Generation IoT Systems*, https://doi.org/10.1007/978-3-032-08784-3_15

cost of medical care. However, the rapid growth of IoMT has faced massive security concerns, where the digital forensics field has emerged as a necessary field to investigate cyber incidents and ensure IoMT systems security.

Digital forensics is the field of forensic science to identify and collect the digital evidence from digital devices and networks for analysis using scientific methods by maintaining the integrity of the evidence and attribute the evidence to the suspect and present the findings in the court of law [1]. As IoMT devices are becoming more connected and part of future networks, the need for effective forensic processes is greater than ever. With the growth in IoMT, there comes an increase in threats, and this mandates sophisticated forensic methods to safeguard patient information and maintain regulatory conformity as the "Health Insurance Portability and Accountability Act (HIPAA,1996)" [2] and "EU-General Data Protection Regulation (EU-GDPR 2016)" [3], and "ISO 27799:2016" [4] have published.

Cybersecurity is now becoming dependent on Artificial Intelligence (AI). AI is helping cybersecurity experts and digital forensics investigators by responding them quickly if a cyberattack occurs as it allows them to track and detect cyberthreats and system vulnerabilities. Hence, digital forensics and artificial intelligence collectively can accelerate and enhance the accuracy of forensic investigations considerably. AI is capable of identifying suspicious behavior that forensic investigators might overlook if it scans a vast amount of data from medical devices. The main points to be discussed in this chapter are the problems, future directions, and the role of AI in supporting digital forensics in the Internet of Medical Things (IoMT).

This chapter's highlights the cybersecurity and digital forensics challenges that IoMT environment presents and how AI innovations have revolutionized this domain and assisting to deal with such cybersecurity challenges. The chapter starts with a general understanding of the Internet of Medical Things (IoMT), its application in healthcare, challenges in IoMT, and cyber threats in IoMT with case studies. The chapter provides an overview of digital forensics in IoMT as it relates to its application of providing security and sustainability in new healthcare systems.

Lastly, the chapter also discusses briefly the evolving trends in IoMT forensics, including use cases and methodologies for AI-driven IoMT forensics, and future directions. By the end of this chapter, readers will learn about the most important developments that are revolutionizing the IoMT field.

2 Understanding the Internet of Medical Things (IoMT)

2.1 Definition and Scope

IoMT is an Internet of Medical Things that comprises systems of medical equipment, healthcare applications, and cloud platforms that store, and analyze data on health [5]. They are remote patient monitoring systems, robotic surgery systems, smart insulin pumps, wearable fitness watches, and pacemakers. IoMT is enabled by wireless

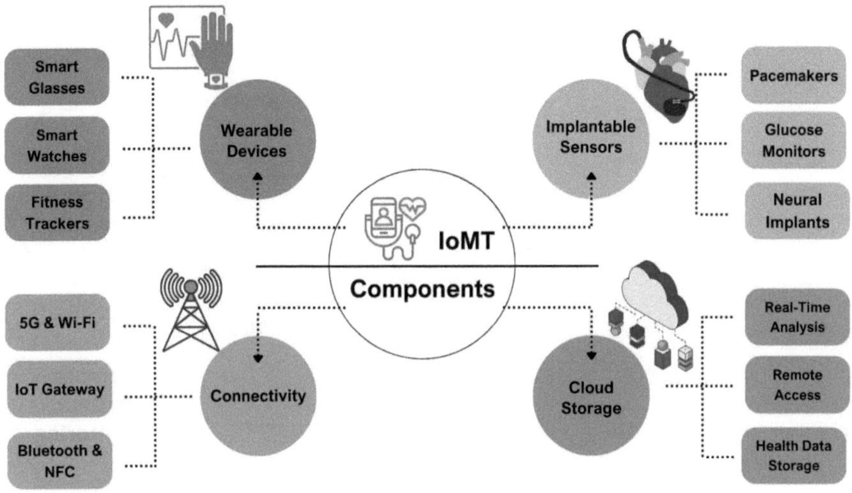

Fig. 1 Key components of the Internet of Medical Things (IoMT)

communication technologies such as cloud computing, Wi-Fi, Bluetooth, and 5G to advance patient care as well as operations efficiency in medical environments. Patient care as well as operations efficiency in medical environments patient care as well as operations efficiency in medical environments. Figure 1 shows the key components of the IoMT environment.

Notwithstanding its benefits, IoMT poses cybersecurity and digital forensic issues. The medical data sensitivity, use of cloud storage, and interconnectivity of devices expose it to cyber threats. Strong security practices, regulatory abidance, and forensic preparedness need to be ensured so that patient information can be safeguarded and faith in IoMT systems is not lost.

2.2 *Applications in Healthcare*

Remote Patient Monitoring

IoMT devices allow the continuous tracking of people suffering from chronic conditions like diabetes, hypertension, and heart conditions. Wearable sensors, smart glucometers, and smart blood pressure machines can offer critical, real-time health information to healthcare providers. Figure 2 shows the various applications of IoMT in healthcare. This enables intervention well within time and reduces the frequency of people requiring hospitalization. It helps patients respond and also has an impact on the reduction in healthcare costs due to emergency admissions.

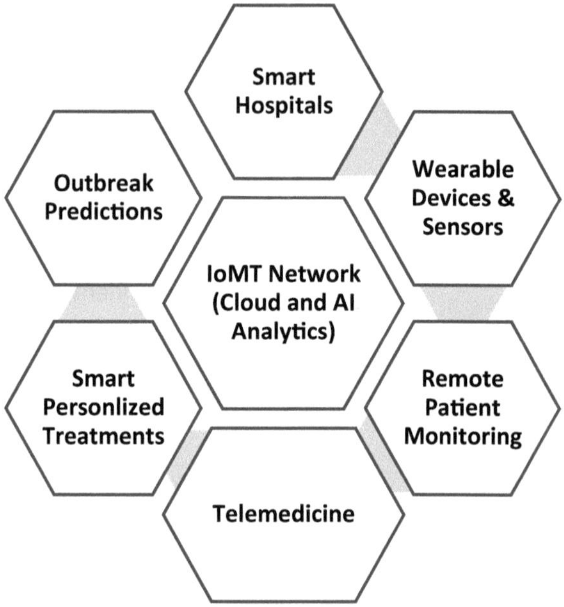

Fig. 2 Applications of IoMT in Healthcare

Telemedicine

IoMT enhances telemedicine because patients are now able to have remote consultations, diagnoses, and prescriptions. With the help of intelligent diagnostic devices, wearable ECG monitors, as well as AI-driven chatbots, doctors are now capable of assessing the patient's condition without the patient having to present in the clinic. This is especially useful to those living in rural settings or underprivileged regions where there is limited access to physicians. Transfer of real-time data further aids physicians in making accurate decisions promptly.

Personalized Medicine

IoMT enables personalized treatment regimens through capturing and analyzing patient-specific data. Asthma smart inhalers, diabetes continuous glucose monitoring, and cloud-connected drug delivery systems guarantee treatment is personalized for an individual. AI analysis offers predictive health risk prediction and proposes prophylactic treatments, enhancing response to therapy and limiting side effects.

Smart Hospitals

IoMT is transforming the idea of hospital operations. Automated inventory management systems monitor medical consumables and medicines, maintaining the ideal inventory levels and minimizing wastage. Radiofrequency identification wristbands make it easier to track patients, with movement tracking and blocking unwanted entry, which simplifies workflow. Predictive equipment maintenance of MRI and ventilators prevent sudden breakdowns and provide continuity of patient care. Predictive equipment maintenance of MRI and ventilators prevent sudden breakdowns and provide continuity of patient care.

2.3 Challenges in Internet of Medical Things (IoMT)

The Internet of Medical Things (IoMT) is very advantageous but it has come across many problems in its large-scale acceptance and adjustment.

Security Vulnerabilities

Most IoMT devices have very weak security features; therefore, hackers can break into them without much trouble. Weak encryption, misconfigurations, out-of-date software, or insecure network connections are some of the ways in which hackers can illegally get access to patient information or even interfere with medical operations. The latest ransomware attacks on hospitals [6] and data breaches are clear signals that there are security loopholes that need to be closed.

Data Privacy Concerns

IoMT devices facilitate the gathering of highly personal health data therefore it is of utmost importance that they follow rules such as "Health Insurance Portability and Accountability Act (HIPAA)" [2] in the U.S. and "EU-General Data Protection Regulation (EU-GDPR)" in Europe [3]. Otherwise, data could be reached by unauthorized persons who might use it improperly or share it in violation of privacy rules, which would lead to an erosion of trust and privacy breach, not to mention the serious legal consequences that could follow.

Interoperability Issues

An Internet of Medical Things (IoMT) ecosystem is a web of healthcare devices from different vendors who are all communicating with their own protocols and using different data formats. The lack of standardization, however, significantly limits the compatibility of devices, health providers, and Electronic Health Record (EHR) systems [7], thereby reducing the efficiency of the whole system. If the interoperability is poor, it can lead to inefficiencies, misinterpretations, and even delays in patient care. The situation, undoubtedly, requires the establishment of common protocols and standards that would result in better cooperation and improved patient outcomes.

Resource Constraints

They are usually small, handheld or even implantable, which means that their computing power, memory, and battery are limited. This also means that they cannot support such high security functions as real-time encryption or AI-driven threat detection without the risk of the performance of the device being affected. Table 1 shows the various data types that are generated by different data sources. The functions of security that are power-efficient and the best kind of firmware are necessary to give

Table 1 Data sources in healthcare

Data source	Data generated	Data type
Electronic health records	Patient medical history, clinical results	Text, numerical
Diagnostic equipment	MRI scans, X-rays, blood tests	Image, text, numerical
Digital wearables	Heart rate, activity data	Text, numerical, image
Healthcare surveys	Research findings, clinical reports	Text, numerical, image

up some functions to gain protection. These problems can be resolved through the joint efforts of healthcare organizations, regulatory bodies, and technology vendors because only they can make IoMT devices interoperable, compliant, and secure without losing efficiency and reliability.

3 Cyber Threats in IoMT Environments

3.1 Major Cyber Threats

IoMT systems are susceptible to numerous cybersecurity threats due to their networked nature and data-centric operations. The increased reliance on network-dependent medical devices exposes healthcare infrastructure to cyberattacks that imperil patient safety, data integrity, and operational efficiency. An in-depth understanding of such threats is important for designing effective security protocols. Table 2 shows common major cybersecurity threats in IoMT and their implications:

Table 2 Key cybersecurity threats in IoMT

Threat	Description
Cybersecurity vulnerabilities	Many IoMT devices lack built-in security measures, making them easy targets for cybercriminals
Data breaches	The sensitive patient information stored in IoMT systems can be stolen and misused
Ransomware attacks	This allows malicious actors to lock down medical devices and hospital networks, demanding ransom payments to restore access
Lack of standardization	The lack of common security standards across various IoMT devices raises compliance issues
Unauthorized access	Hacking can be used to take control of important medical equipment
DDoS attacks	Vulnerable IoMT devices can be utilized in botnets to keep hospital networks down for a long time
Supply chain risks	Such vulnerabilities can be from third-party vendors, impacting device security before deployment

It shows why there should be more rigorous cybersecurity practices in IoMT platforms. Healthcare organizations need to have proactive security controls active, including end-to-end encryption, regular firmware updates, and compliance with standardized security controls, to counter these threats. By safeguarding these vulnerabilities, healthcare professionals can make the IoMT ecosystems more secure and reliable to provide improved patient safety and regulatory compliance.

3.2 *Case Examples*

Real-World IoMT Cybersecurity Incidents

These threats are not merely hypothetical. There have been some well-publicized cybersecurity attacks that have exhibited the weaknesses of IoMT systems. The following case studies document the actual effects of IoMT security breaches. Table 3 summarizes these events and their effect.

Case Study 1: Magellan Health Data Breach (2020)

Magellan Health, an American healthcare organization, was attacked by a cyberattack in 2020 that exposed over 364,000 patients' personal data. It began as a phishing attack that allowed the hackers to get access into the system, plunder sensitive data, and deploy ransomware. The attack illustrates the possibilities of data breaches and ransomware attacks [8].

Case Study 2: WannaCry Ransomware Attack on NHS (2017)

In 2017, WannaCry ransomware hit the UK National Health Service (NHS), encrypting patient records and interfering with emergency services on more than 200,000 computers, including IoMT devices. The malware also caused delays in surgeries. It is a shining example of the risks of ransomware attacks and the necessity for regular updates of security patches to prevent such attacks [9].

Case Study 3: Medtronic Cardiac Device Vulnerability (2019)

Security researchers, in 2019, discovered that Medtronic's implanted cardiac devices, including pacemakers, contained vulnerabilities that enabled hackers to remotely

Table 3 Notable IoMT cybersecurity incidents

Incident	Year	Impact
Magellan health data breach [8]	2020	360,000 patient records exposed
WannaCry attack on NHS [9]	2017	Ransomware disrupted healthcare services
Medtronic cardiac device vulnerability [10]	2019	Exposed pacemakers led to a recall
SingHealth breach [11]	2018	1.5 million patient records compromised
St. Jude medical pacemaker exploit [12]	2017	Security flaws forced an FDA recall

alter device settings. It was a serious enough vulnerability to warrant a recall and security patches to protect the patient. This event points to the dangers of illegal access and supply chain threats [10].

Case Study 4: SingHealth Data Breach (2018)

Singapore's largest healthcare group, SingHealth, was hacked by a colossal data breach in 2018 exposing the health record of 1.5 million patients and even that of Prime Minister Lee Hsien Loong. The hackers took advantage of the system's weak vulnerabilities, shedding sensitive information. The incident illustrates how big a threat data breaches are and how much stronger cybersecurity measures should be enacted [11].

Case Study 5: St. Jude Medical Pacemaker Exploit (2017)

In 2017, security experts found vulnerabilities in St. Jude Medical pacemakers that would allow hackers to drain the battery remotely or change key functions. The news compelled the FDA to order a safety recall and apply security patches. This example illustrates unauthorized access threats and supply chain risk exposure [12].

All these examples require the development of end-to-end security solutions for assurance of security in IoMT infrastructure. Healthcare organizations can ensure the security and reliability of IoMT ecosystems by resolving issues such as the above, thus arriving at improved patient safety and regulatory compliance.

4 Digital Forensics in IoMT

4.1 Why Digital Forensics is Critical

The science of identification, collection, and analyzing the digital evidence while preserving the integrity of digital evidence is known as digital forensics [13]. Digital forensics plays a major role in securing Internet of Medical Things (IoMT) environments by analyzing cyber breaches, ensuring patient safety, and maintaining data integrity. The main goal of IoMT forensics is to find artifacts from medical devices and healthcare networks, and understanding how a cyber incident happened and using that knowledge in court, if required. Because of the particular kinds of devices and data involved, IoMT forensics is not the same as traditional computer forensics. IoMT devices typically have limited memory and processing capacity. They also collect personal information such as blood pressure, heart rate, and prescription details. To protect patient privacy, this data must be handled carefully. An IoMT digital forensics specialist must be able to recover evidence from smart medical devices in a forensically sound manner without altering or erasing it. A proper search and seizure form with chain of custody must be maintained to ensure that the evidence is admissible in court.

One of the main challenges in IoT forensics is the wide variety of devices that are available. It is difficult to develop a forensic tool that works on all devices since they may have different operating systems, storage systems, and communication protocols. Investigators must also be vigilant and proactive to acquire the evidence because some devices might not store data for very long.

There are a number of areas of difference and similarity between traditional forensics and IoMT forensics as shown in Table 4. When it comes to sources of evidence, classical evidence might be computers, mobile phones, servers or gateways. In IoT forensics, the evidence might be domestic appliances, automobiles, tags readers, sensor nodes, medical implants in animals or humans, or other IoT devices.

Investigating Cyberattacks

When an attack happens, digital forensics identifies the cause of the breach, which systems are impacted, and how much damage was done. Through device logs, network traffic, and medical device behavior, forensic analysts can recreate the chain of events that led to an attack.

Ensuring Patient Safety

Infiltration of IoMT devices, like pacemakers or insulin pumps, can be life-risking. Forensic analysis on the digital side detects tampering and unauthorized modification, making medical devices work as planned.

Protecting Data Integrity

Medical records must be genuine and not altered to maintain patient trust and lawfulness. Digital forensics maintains the integrity of patient data by identifying unauthorized changes and preserving electronic evidence for litigation or audit.

Table 4 Differences between traditional and IoMT forensics

Feature	Traditional forensics	IoMT forensics
Devices	Computers, Mobile Phones, Applications	Wearables smart devices, Smart monitors
Operating systems	Standardized (e.g., Windows or MacOS)	Diverse and custom OS
Data types	Media files, Documents, Emails	Vital signs, Health logs
Storage	Large storage	Limited storage
Evidence collection	Manual and tools-based	Complex and device-specific
Network traffic	Predictable	High-frequency, Real time
Power availability	Continuous power	Battery-dependent devices
Legal considerations	Clear laws	Complex patient privacy and jurisdiction issues
Challenges	Large data volume	Various types of devices, Data privacy

4.2 AI-Driven Innovations in Digital Forensics

Artificial Intelligence (AI) offers numerous ways to make digital forensics in the Internet of Medical Things (IoMT) faster and more effective. One of the primary advantages of AI is automation. AI tools can automatically scan logs and medical records to identify signs of a cyberattack. Artificial intelligence is capable of detecting abnormal activities at a very high speed, for instance when someone changes a prescription for a patient without their knowledge.

Manually searches through the databases or logs of the devices with traditional methods can be very time-consuming. AI systems can find suspicious behavior by using machine learning and pattern recognition. Besides that, machine learning models can be employed to train the historical dataset to learn the attack techniques or to predict the new threats. The AI also is able to help in the process of event timeline reconstruction. The investigation timeline is the sequence of events of an incident that constitutes the core of the investigation. Besides, AI is very efficient when it comes to complex medical device networks, as it can examine through different data and correlate it from various devices in order to generate a timeline that logically fits. In addition, AI is utilized in the malware detection process. Most of the IoMT attacks are driven by malware. AI-empowered malware detectors can uncover new types of malwares by drawing from the previous examples. This feature is very handy in the case of "zero-day" attacks, since those are new and no one has seen them before.

4.3 Use Cases for AI-Driven IoMT Forensics

This section discusses a few use cases that show the benefits of the combination of AI and digital forensics for the IoMT field. Therefore, let's say a smart hospital with a network of devices is used to treat patients. Suppose that an intruder manages to access the hospital's network and modifies the infusion pumps' settings, causing the incorrect dosage of medication to be administered. AI software employed to monitor the network notifies the security team if they recognize any suspicious activity. The experts in digital forensics then use artificial intelligence to determine the exact time of attack, what devices were hacked, and how the attacker had gained entry to the system. These apps prove how AI can be extremely useful in detecting and investigating crimes in IoMT. AI identifies what went wrong and prevents similar attacks from happening in the future by learning from previous attacks. Table 5 exhibits the summary of the role of AI and forensic outcome in IoMT Environments.

Use Case 1: AI Identifies Infusion Pump Interference in Smart ICU

A smart ICU that had connected infusion pumps suffered a critical incident where the patient was injected with a medication overdose. An AI-based network monitoring system detected abnormal command patterns that were sent to the pump. AI was used by digital forensic experts to recreate the event, find the fake wireless signal

Table 5 Role of AI and forensic outcome in IoMT Environment

Scenario	AI role	Forensic outcome
Infusion pump tampering	Detects anomalies	Finds timeline and attacker info
Insurance fraud	Identifies fake data	Exposes fraud attempt
Unauthorized access to electronic health record (EHR)	Monitors login behavior	Identifies insider threat
Ransomware in smart ICU	Flags encryption behavior	Traces source of attack
Remote device Hijacking	Detects control anomalies	Reveals command origin
Abnormal vital sign manipulation	Compares data to baseline	Detects health data tampering
Data exfiltration via IoT gateway	Analyzes outbound traffic	Identifies leak path and method

source, and track the attack back to a hospital staff account that was hacked. The security breach was patched, the evidence collected was used to prosecute the case, and the security system was improved with the help of the evidence collected.

Use Case 2: AI Detects Insurance Fraud through Altered Wearable Health Data

An insurance firm noticed an unusual increase of health activity data from a claimant's wearable device. The AI investigation revealed that there were some untruthful alterations in the heart rate and movement data that were not in line with the claimant's medical history. Through digital forensic tools, the authors of the deception were found to be a spoofing app. The revelation not only stopped the fraudulent claim from being paid but also resulted in the perpetrators being brought to justice.

Use Case 3: AI Identifies Ransomware Activity in a Hospital's Smart Network

A hospital's systems were incidentally encrypted due to some unknown cause without the knowledge of the hospital staff. An AI-based malware detection system detected the malicious activity signature as that of a ransomware. To help them find out where the attack was coming from, forensic investigators utilized AI-supported log analysis which pointed them to the phishing email as the entry of the attack. The incident report and evidence helped the hospital put in place recovery measures and also train the employees on cyber threats.

Use Case 4: Recognizing Internal Threats in Electronic Health Record (EHR) Systems

An AI anomaly detection tool noticed a strange pattern of after-hours access to patient EHRs. A probe showed that a nurse was logging into patient data of high-profile patients only for her own entertainment. AI-based behavior analysis aided in tracking the access times, login credentials, and device information. The forensic results were employed in adjusting the access controls and strengthening the privacy compliance.

4.4 *The Role of AI in Digital Forensics*

Artificial Intelligence has been revolutionizing the field of digital forensics. New developments in AI have totally changed the way people analyze and understand digital evidence. The use of AI involves a lot of cutting-edge technologies and methods whereby each one of them is aimed at more productive, precise, and deeper forensic investigations. This part of the article discusses the versatile function of AI in digital forensics and looks at it as a means of data analysis, pattern identification, anomaly detection, and information retrieval.

Smart Forensic Data Processing

AI is definitely a champion when it comes to dealing with large amounts of data electronically and it is a situation somewhat complicated by a forensic environment. It can digest abundant data which have come from a number of sources (e.g., hard drives, clouds, social media) very quickly. This capability means that the job of carrying out the evidence discovery phase is greatly shortened and less manual work for the investigators. AI-powered tools can also combine different data points together, identify time-based sequences, and remove the noise, thus enabling digital forensics professionals to concentrate on the evidence that is of interest to them. Moreover, intelligent automation of this level is indispensable for a case to be solved quickly and efficiently as cybercriminals become more and more sophisticated.

Behavioral Pattern Recognition in Digital Evidence

AI's power to spot regularities in the digital world is making huge changes in how forensic investigators uncover the truth. For instance, if a fraudster is committing a crime, pattern recognition enables the detection of repetition of login attempts, the tracking of data exfiltration routes, and the identification of communication with suspects. Such secrets are deeply buried in vast information and would require human analysts a lot of time to find if it were not for AI. Furthermore, these machine learning algorithms are trained on old cases to establish the necessary habits for new ones. This capability does not only increases the investigations' correctness to a high degree but also enables the smart detection of corruption by behavioral tendencies and the tracking of digital footprints.

Advanced Anomaly Detection for Threat Identification

A prominent instance where AI is involved in the digital forensic field is that it can identify anomalies in a system. An anomaly here means the slightest changes in the system which lie beyond the normal pattern and if those changes are not monitored correctly, it may result in a malicious activity. Such cases are unauthorized data access, manipulation of files, or even uncharacteristic system behavior. Traditional rule-based systems usually can't catch these complex threats but AI algorithms learn from historical data to improve their understanding of what is "normal" behavior. By alerting the analyst about suspicious actions as they happen, AI allows the forensic analyst to be present at the earliest possible moment to respond to the situation, maybe even stop the attack. This feature is the main focus in insider threat detection and after-incident forensic investigations where the need for the fastest decision is the most important condition.

Context-Aware Information Retrieval

AI improves the finding of digital evidence in a way that is not limited to simple keyword searches but also includes context-aware semantic search. In other words, this means that the AI-powered forensic tools can go deeper into the relationships between entities and the intent of the query, thus they can get the relevant information more accurately. To illustrate, instead of providing all files which contain the word "password," AI might highlight those conversations that are most likely to point to a breach or data theft. This filtering in a smart way decisively reduces the noise as well as it increases the demand for investigative focus. Also, the AI is able to check the retrieved items against the other pieces of evidence, which makes the work of the investigator more efficient. In addition, the digital landscapes are evolving exponentially in complexity, so that it becomes imperative for intelligent retrieval to ensure that the case is solved efficiently and timely.

4.5 AI Techniques in Digital Forensics

AI have played a vital role to transform the digital forensics techniques by introducing high-tech tools for searching through complex datasets, detecting anomalies, and automating. The deployment of AI in digital forensics is a big set of tools that include machine learning (ML), deep learning, natural language processing (NLP), and also blockchain. In this section, different AI techniques, the features, the implementation, and the assistance coming from them in the investigations of the computer crimes are discussed.

Machine Learning for Predictive Forensics

Machine learning (ML) is a significant factor in digital forensics as it enables systems to have the power to learn from old data and to become better over time without the need for additional programming only. ML algorithms can be applied to file classification, abnormal behavioral detection, and also to forecasting of threats that may arise based on the incidents which have already been occurred, in a forensic application. For example, ML can be utilized to intercept phishing attacks when it analyzes the email metadata and consequently it recognizes the content patterns. Similarly, it can also become a great assistant during the process of categorizing similar cases, deciding which incidents should be dealt first, and finding the fraudulent activities. Therefore, ML results in a significant decrease in transaction times and also it helps to a more accurate verification of results in the digital forensic analysis.

Deep Learning for Image and Audio Analysis

Deep Learning, a subdivision of ML, is highly effective for unstructured data such as images, videos, and audio files, which are usually digital evidences in forensic investigations. Convolutional Neural Networks (CNNs) are able to recognize the faces, objects, or places that match the crime. In like manner, deep learning networks are capable of running through the video of surveillance cameras in order to track a suspect or obtain the number plate of the car involved. Those tools are also among the devices that enable voice recognition as well as speech-to-text transcription from audio recordings. The ability to find the patterns in the high-dimensional data makes deep learning a powerful ally in the fight against cybercrimes, multimedia frauds, and the cases that involve visual or acoustic digital evidence.

Natural Language Processing (NLP) for Textual Evidence

NLP allows forensic tools to comprehend human language when it is in the form of digital communication. However, the human language found in these digital communications must be meaningfully decoded if the forensic tools are to be able to understand, interpret, and extract information from it. For example, emails, chat logs, social media postings, and documents are some of the places where people can hide their malicious intents, and NLP can be used to examine this textual data to identify possible danger, abusive language, or even the intentions of an insider. Sentiment analysis, keyword extraction, and entity recognition work hand in hand to uncover individuals, motives, and incidents. Besides that, NLP is the biggest helper in the process of creating automated reports and translating the multilingual content. The NLP facilitates converting huge quantities of unorganized text into intelligence that is able to carry out the course of the investigation thus increasing the level and speed of the digital forensic investigations.

4.6 Challenges in IoMT Forensics

Though it is critical, IoMT forensics has numerous challenges that make data collection, analysis, and interpretation difficult. Even though it is vital, IoMT forensics (Internet of Medical Things) is still riddled with a multitude of challenges that make it difficult to collect, analyze, and interpret the data. Such devices covertly make use of unregulated standards that result in inconsistent data formats and with limited interoperability between the systems. Most IoMT devices are highly resource-constrained, thus they have very limited storage and processing capabilities, which automatically restrict the logging of the forensic information. Besides these, the continuous transmission of data and the wireless connection also make the system more exposed to attacks and the loss of information. In addition, the privacy and legal issues related to patient data are multiplying the problems of the forensic investigations because these have to be done very cautiously so as not to violate the rules of "HIPAA" [2] and "EU-GDPR" [3] if they are valid in the region.

Table 6 Types, sources, and forensic values of digital evidence in IoMT systems

Evidence type	Evidence source	Forensic value
Logs	IoMT devices	Disclose unauthorized access attempts
Patient monitoring data	Wearable devices	Provides insights for security incidents
Electronic health records	Cloud storage	Helps reconstruct attacks timeline
Network traffic	IoMT networks	Detects intrusion sources

Data Volatility

Most Internet of Medical Things (IoMT) devices have limited storage capacity, causing them to overwrite data quickly. As a result, obtaining forensic data can be very difficult unless the data is captured in real time.

Encryption Barriers

While encryption enhances security, it also presents challenges for forensic examinations. Powerful encryption methods may hinder forensic teams from gaining access to critical information, necessitating expert decryption methods or court orders.

Heterogeneous Devices

IoMT devices are sourced from different manufacturers, each with its own set of protocols, data formats, and security configurations. This absence of standardization makes forensic data extraction and analysis challenging to perform consistently.

Cloud Storage Issues

Electronic Health Records (EHRs) and device logs for most IoMT systems are stored in the cloud. Judicial authorization, cross-border compliance, and technical competence are normally required to obtain remote data. It may take some time to gather forensic evidence from the cloud, delaying investigation. In order to carry out detailed forensic analysis, experts depend on digital evidence. The major types of evidence, their sources, and their forensic value are shown in Table 6.

By considering these forensic challenges and using major evidence sources, healthcare organizations can increase the effectiveness of their investigation into cyber threats, strengthen security measures, and protect patient information.

5 AI-Driven IoMT Forensic Methodologies

AI plays a crucial role in locating incidents in IoMT ecosystems. It is an automated process that continuously trying to find patterns, logs, and behaviors. The intelligent entities can detect any anomalies in the system without much effort, like if there is a case of unauthorized access or the device behaves abnormally, or if there are some unexpected changes in the patient data. AI, on the other hand, provides a real-time

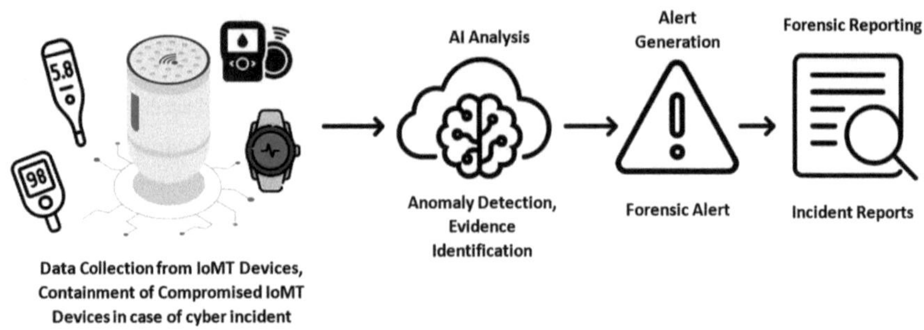

Fig. 3 AI-driven digital forensics process in IoMT system

threat detection that is more effective than the traditional one. Besides, it also collects data from multiple sources and hence can be the earliest to spot the breaches or the devices that are vulnerable. A proactive detection like this is absolutely necessary in healthcare facilities as even a small delay can result in safety risk or loss of data. A conceptual representation of the AI-Driven digital forensics process in IoMT system is illustrated in Fig. 3.

5.1 AI-Driven Incident Response

The instant response to cybersecurity threats in Internet of Medical Things (IoMT) settings is extremely important. Incident response is a systematic approach of threat identification, containment, and elimination that ensures the threat is not allowed to grow into a catastrophe. A response to an incident is a systematic approach which involves detection, isolation, and solving the security issue before it becomes a bigger problem. The healthcare field is a situation in which even a very short period of a delay can be fatal to a patient.

AI assists by facilitating continuous supervision, recognizing threats at their infancy, and implementing automatic countermeasures like device isolation or informing of the staff. Besides, it can even initiate the process of recording evidence right away. Further, AI applications have the capacity of carrying out the threat analysis and ranking the response. This, in turn, makes the decision faster, hence, the damage is less, and patient safety is improved.

5.2 Anomaly Detection in Medical Data Streams

It has been proven that early threat detection can be crucial in preventing cyberattacks. Security Information and Event Management (SIEM) systems are the decision-making networks for IoMT and they follow them continuously for abnormal behavior, such as if there are attempts of access without authorization, if there are data communications going on that are unusual, or if the hardware is failing. SIEM tools get the logs from medical devices, cloud servers, and network endpoints to find out if there is a security breach in real time.

AI is useful in tracking and understanding real-time IoMT data streams to find changes that could be security issues or data manipulation. ML algorithms can be trained with normal behavior of medical devices, so they will be able to highlight any strange movements like those not allowed data access, unexpected turn-offs, or wrong sensor readings just like humans. These changes can be various signs if the device is being tampered or the technical part is not working properly. AI's ability of learning and enhancing its effectiveness continuously enables it to detect even very small threats in complicated IoMT places, thus it can be of great help in making forensic investigation and eliminating risk in healthcare systems on time.

5.3 Containment of Compromised IoMT Devices

It is very important to contain the incident immediately after it has been detected. Containment means that infected IoMT devices are disconnected from the network so that they cannot be used or carry on the attack secretly among other devices. The measures that are taken during containment may include closing communication ports, limit internet access, or restoring affected systems to a safe part of the network. This strategy enables the attack to be lighter while the experts have time to find out where the incident has originated from.

5.4 Contextual Analysis Using Natural Language Processing (NLP)

Natural language processing (NLP) methods are quite efficient in the interpretation of unstructured data that come from medical prescriptions or notes, system logs, or even communication records within healthcare networks. NLP can be a key investigatory tool to extract actions, intents, and to make the events consistent from different sources in patient-device interactions. For example, NLP can point out differences between Electronic Health Records (EHRs) and the device logs that are the actual records, thus fraud or errors can be detected. By understanding the situation of both human and machine-generated text, NLP strengthens forensic analysis in more detailed and diversified IoMT scenarios.

5.5 *Automated Timeline Reconstruction*

AI models are able to reconstruct incident timelines without the intervention of human investigator by going through transaction records, sensor data, and communication patterns among interconnected IoMT devices. In fact, the operation can be considered as revealing of chronology of events just like patient monitoring, device alarms, intrusion attempts, and system failures that enable the visualization of the situation before, during as well as after the wrongdoing. Deep learning models, which have been extensively trained on the past forensic data, can come up with accurate timelines even if they are incomplete or some parts are missing. The act of automating thus not only facilitates the inquiries but also makes it clear for the parties involved to comprehend the scope and the impact of the forensic case.

5.6 *Predictive Forensics for Proactive Healthcare Security*

Healthcare providers are enabled with AI-powered predictive forensics that not only help them to identify the threats but also find the ways of mitigating them. Applying security patches, updating settings, and going over the incident in question one more time to better the plan for the response in the future are all parts of mitigation. These AI systems have a huge accuracy in the recognition of patterns in device behavior, user access, and security incidents from the past. At the same time, AI can come up with preventive measures. This approach is very ideal for a high-risk environment, where the safety of patients is the first and most important issue and also the data should be kept intact. A predictive model-based forensics approach for the Internet of Medical Things (IoMT) security is not only a very important step toward improved privacy but also a kind of proactive forensic readiness.

5.7 *AI-Driven Forensic Acquisition Techniques*

Forensic acquisition is the method of gathering electronic evidence from IoMT devices, network, and cloud storage. Collecting digital evidence in Internet of Medical Things (IoMT) environments can be quite challenging due to the diversity of devices, data confidentiality, and the need for real-time operations. The forensic AI-driven acquisition strategies are designed to facilitate the process by automating the necessary steps, ensuring data security, and efficiently extracting the needed information. These methods are not just breakthrough in keeping the chain of custody and getting the evidence ready for analysis but also in the prevention of the healthcare industry from experiencing any unnecessary disturbances.

Device Identification and Physical Acquisition

AI algorithms continuously scan and also detecting the IoMT appliances that are attached to the different networks. RAM (flash memory) or other storages, physical acquisition, however, can still get the data that is not accessible to software interfaces, like, for instance, logs, firmware, and other information. AI, employing pattern recognition, can assign a device such as heart monitors, insulin pumps, and imaging equipment to a certain category by using metadata and behavioral fingerprints. Such a method not only makes the procedure faster but also more accurate.

Secure Data Extraction from IoMT Devices

AI automates the data extraction process from IoMT devices by launching encrypted and authenticated sessions after it has identified the devices. This privacy assurance along with the integrity of the healthcare data provided is during the process of secure transmission. AI monitors for any abnormal occurrence during extraction, thus it lowers the chances of fraud or tampering of the data and the process of intrusion detection is also continuous.

Data Normalization and Format Unification

IoMT devices are capable of generating data in multiple heterogeneous forms and formats. AI-powered data normalization tools convert raw data into consistent structures so that the interoperability is guaranteed. This particular step uses speech recognition (NLP) and pattern recognition to understand device logs, clinical notes, and structured sensor data in a normalized schema for further analysis.

Evidence Integrity Validation Using Blockchain

To achieve forensic reliability, cryptographic methods like hashing and blockchain-based time-stamping are used to confirm integrity of the evidence. Such methods generate immutable logs, confirm the authenticity of the files, and allow the storage of tamper-evident, which is vital for complying with the legal and regulatory requirements.

Network Forensics and Traffic Acquisition

Network forensic techniques analyze the data packets that are transmitted between IoMT devices, the hospital networks, and the cloud storage. Real-time network traffic capture helps forensic teams detect unauthorized access attempts, malware communications, or exfiltration of data. Network acquisition proves quite useful in detecting advanced persistent threats and unauthorized remote access. AI improves network-level evidence acquisition by connected IoMT devices. It can access packet information, communication records, and device interactions through intelligent sniffers and protocol analyzers. With machine learning, normal and abnormal traffic can be distinguished, and therefore, it allows for the quick and effective gathering of the needed network artifacts.

AI in Cloud Forensics for IoMT Platforms

A vast majority of Internet of Medical Things (IoMT) apps gather and process medical data on cloud platforms. Artificial Intelligence allows cloud forensics to increase the efficiency by spotting virtual resources, identifying compromised containers or cloud APIs, and collecting data from distributed systems without human help. Moreover, it also combines data from various cloud ecosystems, thus providing a single forensic view while still adhering to data sovereignty regulations. The method hereby analyzes remote storage logs, access histories, and backup files to reconstruct security incidents. Cloud-based evidence, however, requires legal permissions, compliance with privacy regulations, and collaboration with cloud service providers.

5.8 AI Integration in Digital Forensic Tools

With the use of AI, computer forensic tools have come to be more efficient than ever. This has especially impacted the way that the investigators carry out the process of discovering, collecting, analyzing, and recording the digital evidence. Traditional forensics tools still rely mostly on manual intervention and have a limited ability to manage small data of a complex nature. AI assists these tools by allowing automation, improving accuracy, reducing the time for analysis, and finding hidden patterns that human detectives are not capable of recognizing. By integrating formalized incident response procedures, advanced forensic acquisition techniques, and forensic domain-specific tools, healthcare organizations are able to boost IoMT security, reduce cyber threats, and assist in the integrity of medical systems.

1. **Automated Triage and Evidence Prioritization**

Automated Forensic tools powered by AI technology can scan computer storage devices and locate files or anomalies that match the search. Machine learning algorithms help decide which evidence is more important by looking for signs of an incident, or examples from the forensic models given. This triage process thus enables the initial steps of investigation to be completed more quickly by focusing on the most relevant data.

2. **Intelligent File and Image Analysis**

Deep learning algorithms are integrated into forensic suites that are used to analyze images, videos, and documents. These forensic tools can spot illicit material or tampered with digital content (such as deepfakes). AI is able to also group visually resembling things and uncover hidden content in media files through steganography.

3. **Natural Language Processing (NLP) for Text and Log Analysis**

Forensic tools powered with Natural Language Processing (NLP) capabilities can perform an unlimited search of textually based evidence, such as emails, chats, and documents. AI is particularly efficient in tasks that require semantic understanding, keyword search, sentiment analysis, and entity recognition. This capability not only aids the investigative team in tracking the context, intent, and the relationship of the parties, but it also makes the work easier and more efficient.

4. **Behavioral and Anomaly Detection**

AI models integrated into network and endpoint forensic tools can also detect abnormal users and systems behavior. Through finding the anomalies, these models can not only detect suspicious activities like privilege escalation, lateral movement, or data theft without permission but also they can be very helpful in incident response and attribution.

5. **Predictive Forensic Analytics**

AI algorithms perform predictive analysis by correlating past case data with patterns revealed in current evidence in order to predict a potential crime scene or even to decide the possible investigation. These hints can go a long way in helping to profile criminals, in tracking cyber threats, or even in court preparation by running simulations of the most likely situations.

6. **Real-Time Monitoring and Live Forensics**

AI-driven technologies allow monitoring and crime investigation with real-time capabilities. They have the opportunity to monitor active sessions, investigate volatile memory (RAM), and gather temporary information in a quick and effective manner. AI acts as a filter and a translator of the live data, recognizing situations which are still active to decide if there are no infringements and to provide a solution if this is the case.

5.8.1 Tools and Platforms for IoT and IoMT Forensics

Some well-known software like Magnet AXIOM uses machine learning and it is able to categorize chat and images automatically to a certain class of things, drugs, guns, and more [14]. Cellebrite Pathfinder utilizes AI to simplify and speed up the decision-making process that is supported by a quick examination of the most important data to promote the search for the investigation of crimes [15].

Maltego monitor live data and employ AI-driven sentiment analysis to identify and evaluate the safety of the community and cyber security incidents that might have occurred or are going on [16]. By way of an example, AI-driven modules can assist in the Optical Character Recognition (OCR) of documents that are scanned, carry out the hash matching in an automatic way, or they decide the similar evidence to be the cluster, hence batch analysis will be easier.

To analyze Internet of Things (IoT) and Internet of Medical Things (IoMT) devices, specialized tools which can effectively handle the great variety of device types, communication protocols, and data formats are essential. Regular digital forensics tools are not always sufficient. That is why the investigators must rely on a combination of general-purpose forensic tools and IoT-specialized platforms.

Table 7 Key forensic tools for IoT and IoMT forensic investigations

Digital forensics tools	Function	Use case in IoMT forensics
Wireshark [17]	Network traffic analysis	Captures and inspects data packets to find out if there is any malicious activity in IoMT networks
Autopsy [18]	An open-source digital forensic analysis	Assists in the retrieval and investigation of forensic information from the storage of IoMT devices
Shodan [19]	IoT device search engine for OSINT	Finds out the unprotected and vulnerable IIoT/IoMT devices
IoT Inspector [20]	Network traffic monitoring	Analyzes IoT network traffic and detects vulnerabilities
Volatility [21]	Extracting volatile evidence	RAM analysis tool for memory dumps
FTK Imager [22]	Acquisition of forensic images	Disk imaging and memory dump tool
Binwalk [23]	Detects malicious firmware elements	Extracts and analyzes embedded firmware

A list of the most productive tools for forensic investigations in the IoT area is given in Table 7.

The AI-powered automation combined with the human forensic methods enables investigators to obtain more efficient security for healthcare institutions and to catch criminals in a shorter time period.

6 Future Directions in IoMT Forensics

With the growth of Internet of Medical Things (IoMT) the cybersecurity issues have also grown. To address these issues an integrated approach involving AI, blockchain, and the development of new legal regulations can be utilized. Blockchain acts as a safe and unchangeable platform for the medical devices' logs, on the other hand, AI can analyze this data and locate the unusual, therefore, it is helping both security and forensic accuracy.

Moreover, machine learning has the ability to carry on the evidence work and thus quickening the moment of reaction to the case. A different approach could be to develop forensic agents that are driven by AI. These utility programs would be installed on IoMT devices. They would continuously check for any suspicious actions, and once they detected any, they would immediately begin gathering the evidence. This would go a long way in cutting down the time of the reaction during the attacks.

The experts in this profession must be taught both the medical technology and AI tools. This will definitely enable the personnel that are prepared to defend the patients in the digital world to be built up. The adoption of AI-powered forensic techniques mandates that the methods be improved so that they can identify new threats, interpret complex medical datasets, and guarantee adherence to healthcare regulations.

1. **Machine Learning and Explainable AI (XAI) Models in Forensics**

Machine learning algorithms improve forensic analysis as they help in pattern and anomaly detection in big data volumes. Traditional forensic processes involve slow, manual log analysis. On the other hand, ML algorithms can process huge volumes of IoMT data by tagging suspicious behavior and correlating events in building cyberattacks. This will make quick and accurate forensic analysis possible to mitigate threats at the earliest point of time. Explainable AI methods would be a feature of AI systems that would be extensively involved in the work of forensic investigators to help them comprehend and trust the results generated by the AI. The openness of the decision-making process would be important not only for the legal admissibility but also for the medical professionals' acceptance.

2. **Federated Learning for Privacy-Preserving Forensics**

In order to ensure that patients' data is protected, federated learning will allow the AI models to carry out the training on the distributed IoMT data without sending sensitive information to the main server. By following this approach, not only do they meet data protection laws, but they also improve their model's robustness.

3. **Real-Time, Edge-Based Forensic Analysis**

Edge computing will give AI algorithms the capability to carry out forensic data gathering and initial analysis in real-time automatically on IoMT devices or at gateways directly with the help of the latest technology. Apart from that, this will also bring about a decrease in the time required for processing, consumption of bandwidth, and dependence on cloud infrastructure.

4. **Integration with Blockchain for Immutable Evidence Logging**

Integration of AI with blockchain will also enable the logging of forensic data and acquisition events that cannot be altered without the record being detected. This combined method will not only secure the chain of custody but also make digital evidence more trustworthy to the court.

5. **Enhanced Anomaly Detection Using Multi-Modal Data Fusion**

AI-enabled systems will get the information from various sources such as sensors, network traffic, and user behavior in order to unearth very subtle anomalies that will be the signs of cyberattacks or device malfunctions; hence, they will not only increase detection rates but also reduce false positives.

6. **Automated Incident Response and Recovery**

AI is going to be more influential in conducting the whole automated incident response, containment, and recovering parts that have been designed for IoMT in the orchestration of the automated sectors. These systems will be able to reduce downtime and the extent of loss by identifying threats with speed and starting counteractions without delay. AI-powered systems can monitor IoMT networks for anomalies such as unusual data transfer and attempts to access the network or devices without permission, or faulty devices continuously. AI also detects threats in real time so that security teams can act before a breach extends. AI system also reduces false positives by discriminating between normal variation in behavior of medical devices and real security events.

7. **Standardization and Regulatory Framework Development**

The rapid increase in the use of AI for the IoMT forensics will lead to the creation of the industry-wide standards and regulatory frameworks that are ethical, legal, and technical best practices. Acceptance and trust will be developed through the adherence to the correct ethical and legal principles.

6.1 Policy Recommendations

For proper containment of evolving forensic challenges, international security standards need to be put in place and encouraged by policymakers and regulatory agencies.

Global Standards for IoMT Security

The absence of common international standards for IoMT security means that there are significant issues, particularly in the case of cyber crimes that occur across borders. Setting up worldwide rules on data encryption, device integrity, and forensic reporting is a matter of utmost importance if we want to secure the medical systems that are connected. The "U.S. Food and Drug Administration (FDA)," the "European Medicines Agency (EMA)," and the "World Health Organization (WHO)" are the organizations that definitely need to be involved if we are to succeed in this. The FDA governs the cybersecurity of medical devices in the U.S. (FDA, 2023) [24], EMA is responsible for the safety of the devices within the EU (EMA, 2023) [25], while WHO provides the global digital health security guidance (WHO, 2022) [26].

Compliance with Medical Data Security Standards

Apart from international organizations implementing similar legislation and medical data protection standards, the adherence to these standards forms the basic foundation for the protection of the Internet of Medical Things (IoMT) environment. The "Health Insurance Portability and Accountability Act (HIPAA)" in the United States of America (USA) defines locally accepted principles for confidential, reliable, and available patients' health records [2]. Besides, the General Data Protection Regulation (GDPR) in the European Union gives a framework for information handling in general, and in particular, for the privacy of the health-related data of individuals [3]. The compliance with these regulations is a must for the legality of the forensic process and the ethical performance of healthcare investigation-related issues.

7 Conclusion

Internet of Medical Things (IoMT) has significantly impacted the healthcare sector to a great extent and is projected to bring along many opportunities and also cyber security challenges. Conventional digital forensics methods that are safe, yet less sensitive, are quite insufficient for the complicated nature of the Internet of Things that are not only large but also far beyond in terms of heat and sensitivity of the medical systems. Still, with the intelligent agent's help, the forensic performance is improved, this, in turn, makes the forensics investigation lighter, more accurate, and more profound. AI serves as a powerful tool, when used as a driving force, enables to unleash new horizons and the abilities to detect anomalies, recognition of objects, predictive forensics, and automate analysis, but at the same time overreliance on AI-driven techniques can also be critical in healthcare environments.

This chapter discusses AI-Driven digital forensics and cyber security issues in the Internet of Medical Things (IoMT). Additionally, it highlights the benefits of fusing AI with forensic features, as well as offers a concise plan for the deployment of AI-powered forensics while complying with international standards, and training skilled personnel, which will be a great step forward in the fight against cyber threats in healthcare.

AI-Driven digital forensics in the IoMT ecosystems would only be effective if there is a collaboration between the technology manufacturers, healthcare facilities, government, and the regulatory authorities. With the appropriate investment in AI-driven forensic systems, security infrastructures, and worldwide regulations, the future of healthcare may be both connected and secure.

References

1. National Institute of Standards and Technology (NIST): Digital Forensics. In: Computer Security Resource Center Glossary. https://csrc.nist.gov/glossary/term/digital_forensics
2. U.S. Department of Health and Human Services (HHS): HIPAA Privacy Rule. In: HHS.gov. https://www.hhs.gov/hipaa/for-professionals/privacy/index.html
3. General Data Protection Regulation (GDPR): In: GDPR.eu. https://gdpr.eu/tag/gdpr/
4. International Organization for Standardization (ISO): ISO 27799:2016 – Health informatics — Information security management in health using ISO/IEC 27002. https://www.iso.org/standard/62777.html
5. Kurtz, J.: Securing Internet-Connected Medical Devices. In: NIST Manufacturing Innovation Blog. https://www.nist.gov/blogs/manufacturing-innovation-blog/securing-internet-connected-medical-devices

6. Whittaker, Z.: How the ransomware attack at Change Healthcare went down: a timeline. In: TechCrunch, https://techcrunch.com/2025/01/27/how-the-ransomware-attack-at-change-healthcare-went-down-a-timeline/
7. Electronic Health Records. In: Centers for Medicare & Medicaid Services. https://www.cms.gov/priorities/key-initiatives/e-health/records
8. Extent of Magellan Health Ransomware Becomes Clear: More Than 364,000 Individuals Affected. In: HIPAA Journal. https://www.hipaajournal.com/extent-of-magellan-health-ransomware-becomes-clear-more-than-364000-individuals-affected/
9. NHS England business continuity management toolkit case study: WannaCry attack. In: NHS England. https://www.england.nhs.uk/long-read/case-study-wannacry-attack/
10. Kovacs, E.: Medtronic releases patches for cardiac device flaws disclosed in 2018, 2019. In: SecurityWeek, https://www.securityweek.com/medtronic-releases-patches-cardiac-device-flaws-disclosed-2018-2019/
11. Tham, I.: Personal info of 1.5m SingHealth patients, including PM Lee, stolen in Singapore's worst cyber attack. In: The Straits Times. https://www.straitstimes.com/singapore/personal-info-of-15m-singhealth-patients-including-pm-lee-stolen-in-singapores-most/
12. Abbott Laboratories Pacemaker Vulnerabilities (Cyber Alert CC-1616). Published 31 Aug 2017. In: NHS Digital. https://digital.nhs.uk/cyber-alerts/2017/cc-1616
13. Digital Forensics. In: National Institute of Standards and Technology (NIST). https://csrc.nist.gov/glossary/term/digital_forensics
14. Magnet Forensics. In: Magnet AXIOM 2.0 – Magnet.AI. https://www.magnetforensics.com/resources/magnet-axiom-2-0-magnet-ai/
15. Pathfinder. In: Cellebrite. https://cellebrite.com/en/pathfinder/
16. Maltego Monitor. In: Maltego Technologies. https://www.maltego.com/monitor/
17. The Wireshark Foundation, https://www.wireshark.org/
18. Autopsy – Digital Forensics, https://www.autopsy.com/
19. Shodan Search Engine, https://www.shodan.io/
20. IoT Inspector, https://inspector.engineering.nyu.edu/
21. The Volatility Foundation, https://volatilityfoundation.org/
22. FTK Imager - Forensic Data Imaging and Preview Solution, https://www.exterro.com/digital-forensics-software/ftk-imager
23. Binwalk, https://www.kali.org/tools/binwalk/
24. Medical Device Safety. In: U.S Food & Drug. https://www.fda.gov/medical-devices/medical-device-safety
25. European Medicines Agency, https://www.ema.europa.eu/en/homepage
26. World Health Organization, https://www.who.int/

Secure RFID Mutual Authentication Protocol Based on Elliptic Curve Cryptography for IoT Edge Environments

Hind Timouhin, Salma Bendaoud, Fatima Amounas, Mourade Azrour, Mohamed Badiy, and Charaf Hamidi

Abstract The Internet of Things (IoT) has revolutionized how smart devices interact, offering a wide array of applications through seamless connectivity. The IoT has become a revolutionary technology, driving the interconnectedness of smart devices across diverse applications. Among the various technologies enabling IoT, Radio Frequency Identification (RFID) plays a pivotal role in enhancing connectivity, automation, and data collection. However, RFID systems face significant security challenges, particularly regarding communication between tags and readers. As IoT environments are inherently resource-constrained, the need for lightweight and secure authentication protocols is paramount. In response, Elliptic Curve Cryptography (ECC)-based authentication protocols have gained prominence for their efficiency and robust security features. Despite existing advancements, security and performance challenges persist in these systems. This paper proposes a novel authentication scheme based on ECC to address these issues more effectively. The findings show that the proposed protocol delivers enhanced security and improved computational efficiency relative to existing solutions. Its performance makes it an ideal candidate for deployment in a wide range of IoT applications requiring enhanced security measures.

H. Timouhin · S. Bendaoud · F. Amounas · M. Azrour (✉) · M. Badiy
IMIA Laboratory, MSIA Team, Faculty of Sciences and Techniques Errachidia, Moulay Ismail University of Meknes, Meknes, Morocco
e-mail: mo.azrour@umi.ac.ma

H. Timouhin
e-mail: h.timouhin@edu.umi.ac.ma

F. Amounas
e-mail: f.amounas@umi.ac.ma

M. Badiy
e-mail: m.badiy@edu.umi.ac.ma

C. Hamidi
Laboratory of Engineering, Faculty of Science, Agadir Sciences University Ibn Zohr, Agadir, Morocco

M. Azrour and A. Alabdulatif (eds.), *AI-Driven Security for Next-Generation IoT Systems*, https://doi.org/10.1007/978-3-032-08784-3_16

Keywords Internet of Things · Authentication protocols · Cryptography · RFID · Security

1 Introduction

During the last decade, the rapid development of Internet of Things (IoT) technologies such as Radio Frequency Identification (RFID), cloud computing, and big data analytics has catalyzed significant transformations across various domains, including smart cities, healthcare, agriculture, and industrial automation [1–9]. The IoT paradigm envisions a seamlessly connected digital environment, enabling communication and interaction among a multitude of heterogeneous devices such as sensors, tags, mobile terminals, and smart objects via the Internet [10–13]. RFID, as a foundational component of IoT infrastructure, plays a vital role in enabling object identification, data collection, and resource tracking. It has found widespread adoption in applications like e-passports, inventory systems, pharmaceutical logistics, and medical monitoring [14–17]. A typical RFID system comprises three core components: the tag (transponder), the reader (interrogator), and a backend server or database. Tags carry unique identifiers and may perform lightweight computations, while readers scan, interpret, and transmit data to the backend system for further processing and authentication. Despite its extensive use, RFID technology remains susceptible to a variety of security threats. Common attack vectors include tag impersonation, traceability, replay attacks, and denial-of-service (DoS). These vulnerabilities highlight the critical need for secure and efficient authentication protocols tailored to the resource constraints inherent to RFID and other IoT components. Traditional methods predominantly based on hash functions and symmetric cryptography, offer efficiency but often fall short in providing robust resistance against active and passive attacks.

Currently, Elliptic Curve Cryptography (ECC) is regarded as a promising alternative for lightweight public-key cryptography, as it offers strong security with reduced key sizes and computational overhead. ECC is particularly well-suited for low-power devices and constrained environments like RFID systems. Several authentication protocols leveraging ECC have been proposed in literature [17–24] to address the limitations of classical approaches; however, many still struggle to achieve the optimal balance between performance, security, and implementation cost. This chapter proposes an improved RFID authentication protocol based on ECC. The scheme is designed to ensure strong security against known attacks, including eavesdropping, impersonation, and replay attacks, while maintaining low computational complexity, making it suitable for deployment in resource-constrained environments such as IoT devices and low-power RFID tags. By leveraging the efficiency of ECC, the protocol offers lightweight cryptographic operations that do not compromise security, enabling secure and efficient communication between tags and readers. Additionally, the protocol incorporates mechanisms for mutual authentication, session key agreement, and resistance to various security threats, thereby enhancing the overall robustness and reliability of RFID systems in diverse applications, including supply chain management, medical devices, and access control systems.

The following sections of this paper are arranged as follows: Sect. 2 outlines the foundational concepts required to support the understanding the proposed protocol. Section 3 details the design of the proposed ECC-based protocol. Section 4 provides a comprehensive security analysis. Section 5 analyzes the performance of the proposed scheme relative to existing solutions. Section 6 concludes the study and suggests directions for future work.

2 Preliminaries

2.1 Elliptic Curve Cryptography

As a modern public-key cryptosystem, ECC relies on the algebraic framework of elliptic curves formulated over finite fields. ECC is increasingly favored for secure communication in limited-resource environments, including RFID, mobile devices, and IoT networks [17]. İn this paper, we focus on elliptic curves over the prime field F_p, given their balance between performance and cryptographic strength.

Let E be an elliptic curve defined over the finite field F_p. It is represented by is the Weierstrass equation

$$y^2 = x^3 + ax + b(\text{mod } p) \tag{1}$$

where $a, b \in F_p$, and the curve must satisfy the non-singularity formulae:

$$4a^3 + 27b^2 \neq 0 \ (\text{mod } p) \tag{2}$$

The collection of points (x, y) that satisfy this equation, together with a unique point at infinity, constitutes an abelian group under a precisely defined addition law.

2.2 Elliptic Curve Discrete Logarithm Problem (ECDLP)

The strength of elliptic curve cryptography is based on the computational hardness of the Elliptic Curve Discrete Logarithm Problem (ECDLP). Given a point $R_1 \in E(F_p)$ and its scalar multiple $R_2 = \alpha R_1$, recovering the scalar α is computationally hard. This one-way property underpins the security of ECC-based systems, distinguishing them from earlier public-key systems such as RSA, which require significantly larger key sizes to achieve equivalent levels of security [25].

The resistance of ECC to known cryptanalytic attacks, combined with its low computational overhead, makes it particularly attractive for use in resource-constrained environments like RFID systems. In this context, ECC offers both security and scalability, enabling the development of efficient authentication protocols tailored to IoT architectures [26].

3 Proposed Protocol

With the growing deployment of RFID systems in sensitive applications, ensuring secure communication has become a critical concern. Authentication mechanisms are at the forefront of safeguarding these systems against unauthorized access and malicious threats. This paper introduces a lightweight and robust authentication protocol designed specifically for RFID environments, leveraging the cryptographic strength of ECC. It is ideally suited for low-power or constrained environments, offering high levels of security while requiring considerably shorter keys than traditional public-key algorithms. By capitalizing on these advantages, the proposed scheme achieves a favorable balance between computational efficiency and high-level security.

Authentication is carried out in two key phases: the initialization phase and the authentication phase. Table 1 details the symbols and notations utilized across the proposed scheme.

- **Authentication Phase**

Table 1 Protocol notations

Symbols	Description
G	Generator point of the elliptic curve
sk_{srv}	Server's private ECC key
P_{srv}	Server's public key: P_{srv} = sk_{srv}G
ID_T	Secret identifier of the tag
u, v	Random nonces generated by the tag
P1	P1 = uG
P2	P2 = vG
α	Hash used for challenge binding
Q	Q = vG + αP_{srv}
Token_T	Obfuscated identity: ID_T $\oplus$ h (v, S1, α, Q)
S1	Prevents replay attacks, sent to Server
R	R = P2 − (sk_{srv}P1), used to recover identity
V	Final digest for verification
V′	Reader-recomputed digest to check authenticity

This phase outlines the mutual authentication procedure between the tag and the server. As depicted in Fig. 1, the proposed protocol proceeds in four main steps. These steps are designed to establish a secure and trustworthy communication channel, verify the identities of both parties, and prevent unauthorized access or impersonation. Each step involves specific cryptographic operations and data exchanges that ensure both the tag and the server can confidently confirm each other's legitimacy before proceeding with further data transmission. This process leverages the robustness of ECC-based cryptography to provide a lightweight yet effective security mechanism suitable for resource-constrained IoT devices.

Step 1: The tag generates two random values from the finite field F_p and computes two elliptic curve points. It derives a temporary secret via a hash function and constructs another point using the server's public key. To ensure anonymity, the tag XORs its identifier with a hash of the randomness, session ID, and curve points. It then transmits the authentication token and the computed points to the reader.

Step 2: In the second step, the reader appends a session identifier to the received message, creating a combined data packet, and then forwards this combined data to the server for further processing. This session identifier serves as a unique reference that helps the server recognize and maintain the context of the ongoing interaction, ensuring continuity and proper handling of the session. By attaching this identifier, the server can accurately associate the message with a specific session, facilitating

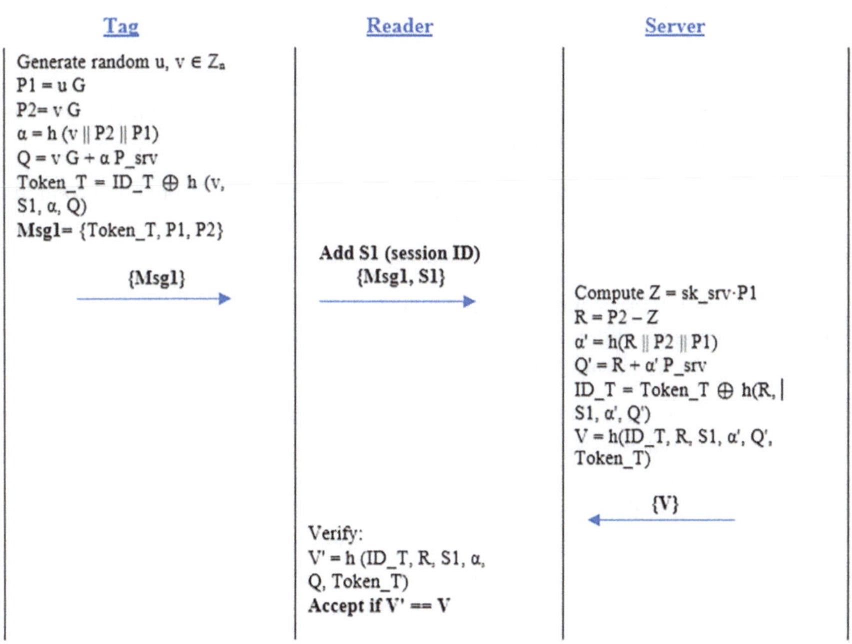

Fig. 1 Proposed authentication scheme

secure and coherent communication flow. This step is crucial in maintaining session integrity and enables the server to perform necessary actions such as validation, response generation, or additional security checks based on the session context.

Step 3: Upon receiving the message, the server computes an intermediate value using its private key and the received elliptic curve point. It reconstructs the tag's randomness and re-derives the corresponding hash and elliptic curve point. Using these values, the server recovers the tag's identifier by reversing the XOR operation. If the identifier is valid, the tag is authenticated, and a verification value is computed and sent to the reader.

Step 4: In the final step, the reader performs its own computation of the verification value using the same parameters. If the value received from the server matches the locally computed one, the authentication is considered successful. If not, the session is aborted.

4 Security Requirements Analysis

Confidentiality: The tag's identity (ID_T) is concealed within Token_T = ID_T $\oplus$ h(v, S1, α, Q), where v is a session-specific random value. An adversary cannot recover ID_T without knowledge of v, S1, or the server's private key sk_{srv}.

Mutual Authentication: The server confirms the tag's identity by reconstructing ID_T from Token_T and verifying V = h (ID_T, R, S1, α', Q', Token_T). The tag authenticates the server by validating V' = h (ID_T, R, S1, α, Q, Token_T), ensuring both parties mutually prove knowledge of secrets (sk_{srv} and ID_T).

Tag Anonymity: Each session generates fresh random values u and v, producing ephemeral parameters P1 = uG, P2 = vG, and a unique Token_T. This prevents adversaries from linking sessions or tracking ID_T.

Integrity: Hash functions ensure tamper detection: α = h (v || P2 || P1) binds critical parameters, while V validates session integrity. Any modification to P1, P2, or Token_T alters α', Q', or V, causing verification failure.

Availability: The protocol's stateless design and lightweight ECC operations minimize computational overhead, enabling faster and more efficient communication without taxing system resources. No persistent secrets are exposed during this process, which significantly reduces potential attack vectors and enhances overall security. This approach prevents adversarial disruption of future sessions by ensuring that even if current session data is compromised, it does not affect the integrity or security of subsequent interactions. As a result, the protocol provides robust, scalable security suitable for environments with limited computational power and high security requirements.

Table 2 Cryptographic operation execution time

Operation	Explanation	Execution time (ms)
T_{db}	Point doubling operation	0.063075
T_{h}	One-way hashing	0.0005
T_{xor}	XOR operation	0.0001
T_{add}	Point addition operation	0.0032

Table 3 Comparison of execution time overheads

Protocols	Total operations	Computation cost (ms)
[27]	$6T_{\text{db}} + 4T_{\text{h}} + 2T_{\text{add}}$	0.387
[28]	$8T_{\text{db}} + 4T_{\text{add}}$	0.517
Proposed protocol	$5T_{\text{db}} + 6T_{\text{h}} + 2T_{\text{XOR}} + 3T_{\text{add}}$	0.328

5 Performance Analysis

This part of the study examines the performance parameters relevant to the proposed mutual authentication protocol, focusing on computational cost. The aim is to evaluate the efficiency and practicality of our protocol in constrained environments, particularly for IoT systems. The computational time cost reflects the processing load during the authentication session between the RFID tag and the server. Table 2 presents the execution times of various cryptographic operations. Given that the bitwise XOR operation has a negligible execution time, it has been excluded from our comparative analysis of computational overhead.

According to Table 3, it can be observed that the proposed protocol achieves the lowest computation cost among all compared schemes, with a total execution time of 0.328 ms. In contrast, Protocol [27] and Protocol [28] require 0.387 ms and 0.517 ms, respectively. This demonstrates a reduction of 16.7% compared to Protocol [27] and 37.5% compared to Protocol [28]. As a result, the findings highlight the proposed protocol's efficiency and low cost, making it well-suited for deployment in IoT-based RFID systems with limited processing and power capabilities.

6 Conclusion

The increasing demand for secure communication in IoT-based environments, particularly in sectors such as healthcare and identity management, highlights the necessity for lightweight yet robust authentication mechanisms. This paper introduced an effective RFID authentication protocol based on ECC, designed to meet the security and performance needs of constrained devices. By leveraging the advantages of ECC, the proposed scheme significantly reduces computational overhead through a minimal number of scalar multiplications, while preserving high levels of security.

Our protocol achieves critical security objectives including mutual authentication, identity confidentiality, and resistance to common attacks. Furthermore, the distribution of computational load favoring the backend server and minimizing processing on the reader ensures suitability for real-time RFID systems. Moreover, the protocol demonstrates strong potential for integration into lightweight applications, balancing performance, cost, and cryptographic strength. These characteristics make it an effective solution for securing RFID-based communications in resource-constrained IoT settings.

References

1. H. Timouhin, F. Amounas, and M. Azrour, "A Survey on RFID Mutual Authentication Protocols Based ECC for Resource-Constrained in IoT Environment," presented at the Lecture Notes in Networks and Systems, 2024, pp. 195–200. https://doi.org/10.1007/978-3-031-48573-2_28.
2. G.-D. Năstase and F. L. Ţiplea, "On a Lightweight Authentication Protocol for RFID," in *International Conference for Information Technology and Communications*, Springer, 2015, pp. 212–225.
3. S. Dargaoui *et al.*, "IoT-Driven Smart Agriculture: Security Issues and Authentication Schemes Classification," in *Proceeding of the International Conference on Connected Objects and Artificial Intelligence (COCIA2024)*, Y. Mejdoub and A. Elamri, Eds., Cham: Springer Nature Switzerland, 2024, pp. 61–66. https://doi.org/10.1007/978-3-031-70411-6_10.
4. N. Ben-Lhachemi, M. Benchrifa, S. Nasrdine, J. Mabrouki, M. Slaoui, and M. ade Azrour, "Effect of IoT Integration in Agricultural Greenhouses," in *Technical and Technological Solutions Towards a Sustainable Society and Circular Economy*, J. Mabrouki and A. Mourade, Eds., Cham: Springer Nature Switzerland, 2024, pp. 435–445. https://doi.org/10.1007/978-3-031-56292-1_35.
5. S. Dargaoui, M. Azrour, A. Allaoui, A. Guezzaz, A. Alabdulatif, and A. Alnajim, "Internet of Things Authentication Protocols: Comparative Study," *CMC*, vol. 79, no. 1, pp. 65–91, 2024, https://doi.org/10.32604/cmc.2024.047625.
6. M. Mohy-eddine, A. Guezzaz, S. Benkirane, and M. Azrour, "IoT-Enabled Smart Agriculture: Security Issues and Applications," in *Artificial Intelligence and Smart Environment: ICAISE'2022*, Springer, 2023, pp. 566–571.
7. C. Hazman, S. Benkirane, A. Guezzaz, M. Azrour, and M. Abdedaime, "Intrusion Detection Framework for IoT-Based Smart Environments Security," in *Artificial Intelligence and Smart Environment: ICAISE'2022*, Springer, 2023, pp. 546–552.
8. C. Hazman, A. Guezzaz, S. Benkirane, and M. Azrour, "Toward an intrusion detection model for IoT-based smart environments," *Multimed Tools Appl*, Aug. 2023, https://doi.org/10.1007/s11042-023-16436-0.
9. C. Hazman, A. Guezzaz, S. Benkirane, and M. Azrour, "A smart model integrating LSTM and XGBoost for improving IoT-enabled smart cities security," *Cluster Comput*, vol. 28, no. 1, p. 70, Nov. 2024, https://doi.org/10.1007/s10586-024-04780-1.
10. S. Dargaoui *et al.*, "Security Issues in Internet of Medical Things," in *Blockchain and Machine Learning for IoT Security*, Chapman and Hall/CRC, 2024, pp. 77–91.
11. S. Dargaoui, M. Azrour, A. El Allaoui, A. Guezzaz, and S. Benkirane, "Authentication in Internet of Things: State of Art," in *Proceedings of the 6th International Conference on Networking, Intelligent Systems & Security*, 2023, pp. 1–6.
12. H. Assmi *et al.*, "A Robust Security Detection Strategy for Next Generation IoT Networks," *CMC*, vol. 82, no. 1, pp. 443–466, 2025, https://doi.org/10.32604/cmc.2024.059047.

13. M. Mohy-eddine, A. Guezzaz, S. Benkirane, and M. Azrour, "Malicious detection model with artificial neural network in IoT-based smart farming security," *Cluster Computing*, pp. 1–16, 2024.
14. U. Iqbal, A. Tandon, S. Gupta, A. R. Yadav, R. Neware, and F. W. Gelana, "A novel secure authentication protocol for IoT and cloud servers," *Wireless Communications and Mobile Computing*, vol. 2022, no. 1, p. 7707543, 2022.
15. P. Bagga, A. K. Das, M. Wazid, J. J. Rodrigues, and Y. Park, "Authentication protocols in internet of vehicles: Taxonomy, analysis, and challenges," *IEEE Access*, vol. 8, pp. 54314–54344, 2020.
16. D. Noori, H. Shakeri, and M. Niazi Torshiz, "Scalable, efficient, and secure RFID with elliptic curve cryptosystem for Internet of Things in healthcare environment," *EURASIP Journal on Information Security*, vol. 2020, no. 1, p. 13, 2020.
17. S. Dargaoui, M. Azrour, A. El, A. Guezzaz, A. Alabdulatif, and S. Ahmad, "ECC-Based Anonymous and Multi-factor Authentication Scheme for IoT Environment," vol. 21, no. 1, 2025.
18. M. R. Servati and M. Safkhani, "ECCbAS: An ECC based authentication scheme for healthcare IoT systems," *Pervasive and Mobile Computing*, vol. 90, p. 101753, 2023.
19. M. Benssalah, I. Sarah, and K. Drouiche, "An efficient RFID authentication scheme based on elliptic curve cryptography for Internet of Things," *Wireless Personal Communications*, vol. 117, no. 3, pp. 2513–2539, 2021.
20. H. Timouhin, F. Amounas, and M. Badiy, "Efficient ECC-based RFID authentication for enhanced IoT security," in *Recent Advances in Internet of Things Security*, CRC Press, pp. 39–46.
21. N. Dinarvand and H. Barati, "An efficient and secure RFID authentication protocol using elliptic curve cryptography," *Wireless Networks*, vol. 25, no. 1, pp. 415–428, 2019.
22. H. Timouhin, F. Amounas, and S. Bouarafa, "Faster RFID Authentication Scheme Based on ECC for Improving the Security in IoT Environment," in *The International Conference on Artificial Intelligence and Smart Environment*, Springer, 2022, pp. 433–439.
23. S. Xie, F. Zhang, and R. Cheng, "Security enhanced RFID authentication protocols for healthcare environment," *Wireless Personal Communications*, vol. 117, pp. 71–86, 2021.
24. H. Timouhin, F. Amounas, and M. Azrour, "New ECC-Based IoT Authentication Protocol for Securing RFID Systems," *SN COMPUT. SCI.*, vol. 4, no. 6, p. 785, Oct. 2023, https://doi.org/10.1007/s42979-023-02220-2.
25. A. Menezes, "The discrete logarithm problem," in *Elliptic Curve Public Key Cryptosystems*, Springer, 1993, pp. 49–59.
26. S. W. Gebregiyorgis, "Algorithms for the elliptic curve discrete logarithm and the approximate common divisor problem," PhD Thesis, University of Auckland, 2016.
27. M. Naeem, S. A. Chaudhry, K. Mahmood, M. Karuppiah, and S. Kumari, "A scalable and secure RFID mutual authentication protocol using ECC for Internet of Things," *International Journal of Communication Systems*, p. e3906, Jan. 2019, https://doi.org/10.1002/dac.3906.
28. S. Gabsi, Y. Kortli, V. Beroulle, Y. Kieffer, A. Alasiry, and B. Hamdi, "Novel ECC-based RFID mutual authentication protocol for emerging IoT applications," *IEEE Access*, vol. 9, pp. 130895–130913, 2021.

MIX
Papier aus verantwortungsvollen Quellen
Paper from responsible sources
FSC® C105338

If you have any concerns about our products,
you can contact us on
ProductSafety@springernature.com

In case Publisher is established outside the EU,
the EU authorized representative is:
Springer Nature Customer Service Center GmbH
Europaplatz 3, 69115 Heidelberg, Germany

Printed by Libri Plureos GmbH
in Hamburg, Germany